Supporting University Ventures in Nanotechnology, Biomaterials and Magnetic Sensing Applications

Julian Peña Dopazo · Fatima Zivic
Editors

Supporting University Ventures in Nanotechnology, Biomaterials and Magnetic Sensing Applications

Policies, Practices and Future

Co-funded by the
Tempus Programme
of the European Union

WIMB

Springer

Editors
Julian Peña Dopazo
Department of Finance
Endor Nanotechnologies, Parc Científic de Barcelona
Barcelona
Spain

Fatima Zivic
Faculty of Engineering
University of Kragujevac
Kragujevac
Serbia

ISBN 978-3-319-61236-2 ISBN 978-3-319-61237-9 (eBook)
DOI 10.1007/978-3-319-61237-9

Library of Congress Control Number: 2017945262

Printed on acid-free paper

This Springer imprint is published by Springer Nature
The registered company is Springer International Publishing AG
The registered company address is: Gewerbestrasse 11, 6330 Cham, Switzerland

Preface

Modern science is a discovery as well as an invention

—J.L. Heilbron

Progress in science can emerge in many different ways. Even when a research project, a technical development, or a scientific theory is rejected or replaced, progress is present in the sense of the refinement of the path in the quest for wider and more accurate knowledge and understanding of nature.

A traditional debate juxtaposing basic research—the fundamental scientific activity not intended to achieve a specific practical purpose—and applied research—the scientific activity aimed at producing tangible results—was once ongoing within the scientific community because the two types of research were pursued in spheres and often not interlinked to each other, even in cases when they should have been. However, interdisciplinarity has always existed in science, though the speed of knowledge exchange pertinent to the dissemination of progress has been limited by the available means of communication at any given moment. An illustrative example could be the development of NASA's—and the equivalent former Soviet Union's apparatus—vast space programme in the closing decades of the twentieth century: undoubtedly, innumerable contributions from many scientific disciplines—from basic to applied—were needed to finally accomplish one of the oldest dreams of humanity: reaching outer space. Or, going back forty-six centuries in time, one can also consider the construction of the Giza Pyramids; such a magnificent result can only be envisaged with the participation of the most advanced science (innovative engineering, new materials, evolved communications… even applied astronomy) of that era, and surely with highly qualified and motivated project managers and technical planners. In both cases, of course, a major injection of financial support had to be made and administrated from different coordinated origins with a common final objective.

Nowadays, most scientists agree in the need of sustaining basic research as a source of potential new applications for Science and the final recipients of the results, being individuals, organizations or Society as a whole. Furthermore, it is commonly assumed that—due to the intrinsic complexity, increasing specialization and specific features of each of the stages related to a scientific development—the process has to be fragmented and performed by different stakeholders in order to become more efficient. *Efficiency* in Science means minimizing the resources needed to get a result and having the vision to keep the project under way if some collateral discovery could arise. So, in terms of scientific management, productivity as a straight measure of efficiency has to commune with the capability to anticipate opportunities. And specialization appears to be a natural process to extend the scientific impact to the very last of the aspects of the world. Any of the general branches in science is currently divided in a myriad of subcategories that form a very intricate system indeed, but this is not a negative fact as far as it *forces* researchers to focus on a very specific field of discovery and try to sort out the very last of its hidden secrets—or find out new ones to be studied—in an iterated and cumulative process. This, in opposite to previous periods when scientists used to work on wider areas of knowledge at the same time, is only possible when the critical mass of dedicated staff is large enough and the resources devoted to the progress of science are sufficient. Even the most recent discipline in applied science, Nanotechnology, has already deployed the usual range in subdivisions of activities: nanomedicine, nanochemistry, nanomaterials, nanoelectronics, nanophotonics, nanometrology. Thus, the research system is developed by diverse players that go from individuals to the biggest companies, governments or even transnational organizations, creating a global net of synergies that boost research, both basic and applied, to a point never known before. It is in this rich discovery and innovation ecosystem that the different stakeholders match their role in the big picture regarding their dimension, purpose, sustainability, objectives and interests. Moreover, the instant access to gigantic online contents on a given research provides the mechanism to facilitate the complementarity of research lines and projects: it becomes possible to reach a global view on the state of the art of a subject and check whether a scientific development is expected to contribute to the progress in that specific matter prior to starting the project.

Let us give an example related to the content of this book to illustrate the interaction among activities inside medical research that has been a standard for the last 50 years: the drug discovery process.

It takes 10–15 years for a compound to reach the market in the form of a new drug, after several screening stages—each one of them perfectly regulated and controlled—that refuses 10.000 former candidates as an average. This process involves all the range of scientific possible scientific actors from most basic research to final marketers and needs of a total investment of around 1 billion euros. (DiMasi JA, Grabowski HG. R&D Costs and returns to new drug development: a review of the evidence. In The Oxford Handbook of The Economics of the Biopharmaceutical Industry. eds. PM Danzon and S Nicholson, 2012, chapter 2:21–46. Oxford, UK: Oxford University Press). The number of novel new drugs approval by FDA averages 25 per year in the period 2005–2013 (US Food and Drug Administration, Center for Drug Evaluation and Research, Novel New Drugs 2014 Summary, 2015). General schematic of time range related to drug discovery

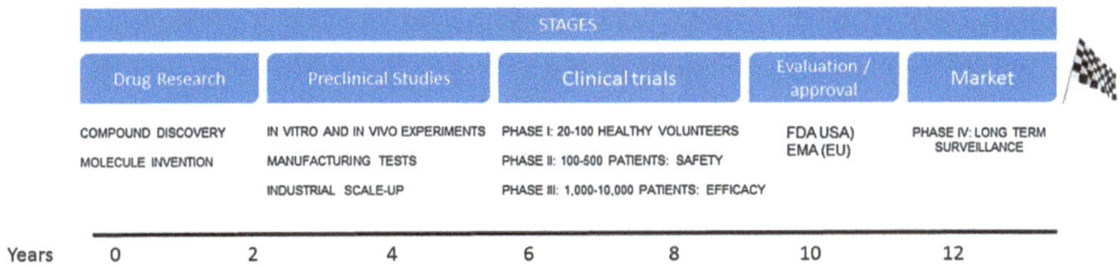

Fig. 1 Drug discovery stages *Image* Endor Nanotechnologies

and use in the market is given in Fig. 1, according to practical experience by Endor Nanotechnologies, company that has business precisely in the area of drug discovery.

The main stakeholders in the drug discovery value chain take their role in the play where they can contribute properly:

- Academia. The general source for the discovery and invention of compounds. A continuous feed of new development possibilities. At this stage, the market is several years ahead, so the scope of the projects relies on excellence in science and widening the technical frontiers beyond current limits.
- Universities. The organizations that bring academic discoveries to a formal stage. Usually, first intellectual property protection for basic developments is carried out at Universities. When a technology transfer office is present, not only IP registration and follow-up are done but also commercial negotiations on developments suitable for exploitation by the rest of organizations involved in the drug delivery process. In this case, the interest for the industry is gaining early access to potential disruptive compounds at a lesser cost compared to more advanced stages.
- Spin-offs/start-ups: Both concepts are applied in many different ways and variations (university spin-off, tech-based start-up, J.M Beraza Garmendia/ A. Rodríguez Castellanos), and currently in a much vaster and more flexible sense. For the scope of this introduction, the assumption is that the act of 'starting up' is wider and inclusive of 'spinning-off' (this is, *starting up* from an already existing organization) and, at last, the role and place in the value chain is equal not regarding nomenclature. Thus, here the term 'start-up' tries to sum up complementary concepts: a university-born entrepreneurial initiative to take benefit of an advanced knowledge representing a clear innovation, with a critical potential impact in the market, often in the initial form of a micro-small private company. The typical role of this kind of start-up company in the drug development process is completing the preclinical and/or Phase I of the development.
- Investors, funders. Regarding the project stage, there are different denominations for the funding partakers: *Seed Capital, Business Angels, Venture Capital*... They bring financial support and often business counselling from initial start-up investments to the next expansion financing rounds up to the first public offering of shares. These specialized professionals used to have a

peripheral vision of the whole sector they invest in and can boost synergies in several areas while participating in the decision boards of their invested companies. Crowd funding is a peculiar new way of reaching financial resources from granular small contributions instead of a one-time, one-source disbursement, and specific online platforms for drug development funding have emerged recently.

- Contract research organizations (CROs). CROs play a central role in the whole chain of the drug development. They usually complement the areas of activities that are not in the core of the research company and can be outsourced to these specialised organizations: regulatory affairs, industrial scalation, clinical trials, market studies; they can collaborate with any of the other stakeholders involved in the development.
- National/Transnational Official Agencies. Any new drug going into the market has to be approved for human use by the official agency that holds the authority over the drug commercialization. The process for approval may vary in every country, but the aim is to keep a close control on the marketed products. This is inclusive of potential long-term undesired side effects, toxicology surveillance and alert protocols in the case of a serious issue affecting patients. The two main agencies that lead the regulatory affairs worldwide are the FDA (Food and Drug Administration) in the USA and the EMA (European Medicine Agency) in the European Union.
- Pharma companies/industry. The pharmaceutical market holds one of the highest volume in sales worldwide. In 2011, the global revenues for this market surpassed $1.000 billion for the first time. The industry is not only interested in brand new developments but also reprofiling and repositioning, pertaining to the different further applications of drugs initially developed for a specific different therapy. Moreover, among pharmaceutical companies providing generic drugs (those with expired patents and thus available for open manufacturing and selling), the business opportunity is still heavily present. A pharmaceutical company can become a licensee in any of the stages of the development, depending on the degree of innovation, effectiveness or business potential can be seen.

On a very different way, the discovery of graphene in 2004 at a university laboratory (Electric Field Effect in Atomically Thin Carbon Films. K.S. Novoselov, A.K. Geim, S.V. Morozov, D. Jiang, Y. Zhang, S.V. Dubonos, I.V. Grigorieva, A.A. Firsov, 2004) was not exactly what we could call a standard process. It is indeed a good example of how basic research on a novel material that started appearing in scientific literature by 1947 in a conceptual level (The Band Theory of Graphite Phys. Rev. 71, 622. P. R. Wallace. 1947) becomes one of the most promising industrial applications beyond fundamental nanotechnology. So promising that the European Union launched its *biggest ever research initiative* called Graphene Flagship in 2014 involving 142 partners in 23 countries by 2015. The consortium behind the organization is, again, composed by stakeholders from university to big industries, including start-ups that have just been founded for the production and processing of graphene itself.

So, it is a general feature that in any activity concerning discovery, new applications and innovation a start-up niche appears. Thus, we should consider what is the role and the reason for a start-up to be created and why is it such a common thing to happen in the university environment. Start-ups aim at transforming conceptual research and basic development into applied innovation. They are the links that allow university research groups to step into the market, since as far as usually, regarding local dispositions, a university itself cannot act as direct promoter, founder or shareholder of an independent private company. Additionally, start-ups can bring to light the rich flow of intellectual knowledge originated in public organizations which are usually left on standby for an applied advance. The dimensional advantage from this point of view is crucial—no big investment nor infrastructure is needed to start-up—just one single potential innovation, a balanced combination of will and intelligence and, possibly, a little bit of good luck, or let us say 'opportunity'. The key point in this know-how transfer system to expand is the high iteration speed: as far as in a start-up ecosystem a business idea can become an incorporated company in a matter of days, this helped by the favourable environment given by Universities and Science Parks themselves. The same ecosystem is quick and efficient in *clearing* start-ups that do not find a demanding match for their offer: the failure rate is very sharp among start-ups. It is a high-risk bet and a very volatile environment. If repeated time after time, once in a while a start-up succeeds in going beyond the survival point, growing rapidly and turning into, or being sold to a big company and an example, as well as an incentive, for new similar initiatives. Because when a start-up succeeds, it is a really big deal. These are just a few of the several so-called *unicorns* in the recent years, start-ups that have achieved a post-money valuation of at least 1 billion dollars. The post-money valuation is the value of the company after a new private investment round or at the completion of the Initial Public Offer—IPO—at the stock market. It fixes the number of equity shares gained by every shareholder after the increase of capital (Table 1).

One of the singularities concerning start-ups is the way of financing the first steps into autonomous activity. This is usually the wage for the first employee and the very least operative expenses concerning basic administration. No traditional financing formula from banks can easily be achieved, regarding two main reasons: a high-risk business plan and no historical backup from the lender's point of view. At this stage, funds support from other organizations rather than banks is crucial. Company Incubators, Science Parks, Technology Transfer Offices, Public Institutions, Private Investors (Seed Capital, Business Angels, Venture Capital, Big Companies, etc.) play a fundamental role in keeping the source of new start-ups flowing. And the shorter the way from the business idea to the market application, and the bigger the market, the easier and faster financial resources can be achieved and multiplied.

This particular aspect—the importance of the expected time to market—can be observed in the profile of the projects granted by the European Union through the Frame Programmes for Research, Development an Innovation and its evolution.

Table 1 *Unicorns* from different countries and industries, listed by the year of foundation

Company	Area of activity	Year of foundation	Country of origin
Vertex Pharmaceuticals	Pharmaceutical	1989	USA
Avast	Software	1991	Czech Republic
Skyscanner	Travel	2001	UK
Portola Pharmaceuticals	Biopharmaceutical	2003	USA
BlaBlaCar	Transportation	2004	France
Mascoma	Biofuel	2005	USA
Spotify	Music streaming	2006	Sweden
Adyen	Financial services	2006	The Netherlands
Opfthotech corporation	Biopharmaceutical	2007	USA
Flipkart	E-commerce	2007	India
Dropbox	File hosting	2007	USA
Avito.ru	E-commerce	2007	Russia
Airbnb	Accommodation	2008	USA
Uber	Transportation	2009	USA
Kik	Communications	2009	Canada
Xiaomi	Electronics	2010	China
Ultragenyx Pharmaceutical	Biotechnology	2010	USA
Delivery Hero	Catering	2011	Germany
Global Fashion Group	E-commerce	2014	Luxemburg

The *7th Framework Programme for Research and Technological Development* that lasted from 2007 to 2013, assigned € 50.521 million assignated and included five general pillars: Cooperation, Ideas, People, Capacities and Nuclear Research. Mainly based on collaborative projects to join academia, research groups and industry to reinforce the strengths of the European Union on the topics that were considered of the major importance at the time the Frame Programme was designed. For the new Frame Programme, Horizon 2020, that will grant € 80.000 million and last from 2014 to 2020, directives show that the general aim is to grant projects that are closer to market than in the previous Frame Programme, using a specific indicator, the *Technology Readiness Levels*—TRL—to measure the distance between a development and the real application:

- TRL 1—basic principles observed
- TRL 2—technology concept formulated
- TRL 3—experimental proof of concept
- TRL 4—technology validated in laboratory
- TRL 5—technology validated in relevant environment
- TRL 6—technology demonstrated in relevant environment
- TRL 7—system prototype demonstration in operational environment
- TRL 8—system complete and qualified
- TRL 9—actual system proven in operational environment

Table 2 Horizon 2020 structure

Main pillars	Excellent science	Industrial leadership	Societal challenges
Areas	• European Research Council • Future and Emerging Technologies • Marie Sklodowska-Curie Actions • Research Infrastructures	• LEIT, Leadeship in enabling and industrial technologies (ICT, Nano, new materials, Biotechnology and Space) • Acces to Risk Finance • Innovation in SMEs	• Health • Food • Energy • Transport • Climate • Inclusive Societies • Security
Transversal programmes	Spreading excellence Science for society European Institute of Innovation and Technology—EIT Joint Research Centre—JRC Euratom		
Industrial pilot (2015–2016)		Fast track to innovation	

Source European Commission H2020 website

What is more important, for the first time single Small and Medium Enterprises (SME), also including start-ups, can apply for individual grants to get funds for their projects in a competitive scheme with the rest of applicants. This tells us about the priority that the European policies bestow to the applied innovation and the kind of incentives to modulate the whole process of technological transfer to the industry. Just to mention the current pillars, Horizon 2020 deploys Excellent Science, Industrial Leadership and Societal Challenges (Table 2).

In conclusion: big companies willing to carry on with innovative projects cannot afford to be present in every little hidden corner of scientific activity, and in this niche is where small start-up companies arise and contribute with cutting-edge developments potentially interesting for the industry. This book is about that particular form of progress in Science.

This book is the result of the contributions from several experts and professionals in the matter, from academia to business managers and entrepreneurs. The focus is on practical aspects of the start-up role, conception and execution from a university environment, though many of the topics can be applied to other fields. We are including articles about some of the desirable features that a scientist that becomes an entrepreneur should have: creativity, flexibility, interdisciplinarity, simplicity and patience, along with willingness and knowledge to adopt innovation. The book covers the areas pertaining to the transfer of technology from research to society. When transferring technology, two main ways may arise: licensing out the further exploitation rights straight from the university departments to the industry and go together with the technology step beyond into market applications for products abd services development, through spin-off and start-up companies. This book has focused on the second process and stakeholders involved, and with several study cases from real life.

Profiles of research entrepreneurs are described, along with categories and the general characteristics of entrepreneurial infrastructure. Different phases of launching university ventures are presented, as well as currently perceived technology transfer systems, important practical considerations for Intellectual Property (IP) protection are included: legal and regulatory affairs, as part of the formal requirements for a business to adopt, complemented with IP rights and strategy aimed to protect and strengthen the know-how. Some aspects of new technologies and their impact on new entrepreneur activities and frameworks are considered since advancements in ICT and nanotechnologies have provided the most significant influence on rapid growth of start-ups nowadays. Within university environment, collaboration with industry and business sector is of the utmost importance for boosting entrepreneurship among young people and researchers to tranfer the technology to the market. New ICT solutions as a support to companies, especially small and medium enterprises (SMEs) can boost competitiveness.

Innovation towards real market product is especially important in nanotechnology where newly generated knowledge is still in laboratory stages, and many new issues can arise related to real applications and company foundation. However, successful stories of starting up a company in nanotechnology sector exist already and this is the novelty making our time. We believe that, beside general theoretical chapters, these successful stories about companies that were founded and do business based on research results are very important for young researchers and professionals willing to start up a tech-based company in the university environment, to motivate them and showcase the possibilities. Accordingly, real business case studies of research transfer and real examples of stakeholders in the system for several successful companies, related to nanotechnology, biomaterials and magnetic sensing applications, as well as practical guide on IP protection, are given at the final chapters.

This work is the result of joint efforts of different companies, academic, and research institutions participating in the large international project, WIMB Tempus, 543898-TEMPUS-1-2013-1-ES-TEMPUS-JPHES, "Development of Sustainable Interrelations between Education, Research and Innovation at WBC Universities in Nanotechnologies and Advanced Materials where innovation Means Business", cofunded by the Tempus Programme of the European Union.

Barcelona, Spain Julian Peña Dopazo

*The original version of the book was revised:
Incorrect volume editor name has been updated.
The erratum to the book is available at
10.1007/978-3-319-61237-9_12*

Contents

Part I
A Framework for Establishing University Ventures

Profiles of Entrepreneurs and Motives for Starting a Business

Vlado Medaković and Srđan Vasković

Abstract In the turbulent environment of constant changes and influences, economic, technologically-technical and social and international factors, we can say that at the beginning of the 21st century we live in the age of entrepreneurship, i.e. in the time of small and medium-sized enterprises, which are readier and more capable of reacting faster to market changes. The theories on enterprise and entrepreneurship define the enterprise and entrepreneurship as immanent phenomena, as two inseparable entities of one body. In that, entrepreneurship implies the directing of resources in the area of their optimal use, while entrepreneur implies the person who acts, who gives concrete forms to jobs, who anticipates risk, create new forms of jobs, expands employment, and enables better and more efficient organization of enterprises capable of meeting all the challenges put in front of it. The role of entrepreneurship is universal. It includes almost all aspects of human life. Every society uses entrepreneurs' experience in management, support and success encouragement, resource activation, motivation and risk awarding, business efficiency, stability and growth, taking responsibility and business risk. Entrepreneurs see their chances where others see confusion and chaos. Key players enabling such changes are free entrepreneurs, young entrepreneurial managers, persistent innovators, and creators of a new business world. Entrepreneurs make decisions to form new enterprises on the basis of their personal, subjective motives. Regardless of what motivates them, entrepreneurs assume an obligation towards a business idea, i.e. a project, and, by means of that, they also dictate the future success of an enterprise. Empirical research shows that there is potential among students of engineering disciplines, that is only necessary to be adequately mainstreamed through the educational programs for entrepreneurship.

V. Medaković (✉) · S. Vasković
Faculty of Mechanical Engineering, University of East Sarajevo, Vuka Karadzica 30, 71123 East Sarajevo, Republic of Srpska, Bosnia and Herzegovina
e-mail: vlado.medakovic@gmail.com

S. Vasković
e-mail: srdjan_vaskovic@yahoo.com

J. Peña and F. Zivic (eds.), *Supporting University Ventures in Nanotechnology, Biomaterials and Magnetic Sensing Applications*,
DOI 10.1007/978-3-319-61237-9_1

Keywords Entrepreneurship · Entrepreneur · Business · Profiles of entrepreneurs · Motives of entrepreneurs

1 Introduction

Today, in great numbers, people see a future in entrepreneurship and have the wish to run their own business. Some of them are motivated by a wish to provide themselves a job or to do a kind of job that they enjoy without a supervising boss. Some others feel a need for a greater challenge or for better working conditions. Many people search for activities that will make them financially independent. The reasons for creating one's own business are numerous. Such reasons depend on personalities. Not everyone is capable of running a business or grasps what is necessary to run a certain project. This suggests that there is a great number of those who may become entrepreneurs, who have entrepreneurial characteristics and who could express their potential in an effective way. Entrepreneurship has been in focus during the previous years, due to the significant benefits realized by small and medium-sized enterprises in the overall development of national economies of developed countries. Entrepreneurship is a multidimensional phenomenon which is involved in many disciplines. The entrepreneurship theories come from various schools, with different opinions, and many authors identify various different factors that lead to entrepreneurial ventures (Kalyani and Kumar 2011). The key word that circulates around the economy development in numerous countries is education. The emphasis of entrepreneurship and entrepreneurial education is on starting a business and the growth of entrepreneurial activities, and especially on the occurrence of business problems. Enabling the young through the education, i.e. the introduction of entrepreneurship in the curricula within the present educational system—starting from the lowest educational level—creates more possibilities for a continuous increase in awareness and expands the horizons to future entrepreneurs. The main goal of entrepreneurial education is to provide the interested participants with adequate knowledge, skills and the abilities to engage in entrepreneurial ventures in various sectors (Kalyani and Kumar 2011).

Over recent decades, the change of economic structure on the global level has been characterized by a redistribution of those employed in the economic sector, where an overflow of labour force from the primary and secondary sector to the tertiary sector has occurred. The valorisation of real (hidden) potentials and productive employment have been greatly conditioned by the affirmation of entrepreneurial behaviour, starting from family, small and medium-sized enterprises and progressing to a higher (aggregate) level. A multitude of factors affect the processes of intensifying the development of entrepreneurship (including, among others, market, financial, urban, communal, legal, and human factors), which are the ambience in which the entrepreneurship is affirmed as an efficient way of running and developing a business, and also the improvement of quality of life and work in certain environments (Melović et al. 2011).

Small and medium-sized enterprises are one of the key sectors of the economy development in European countries. They are significant initiators of innovations, employment and social and local integration in Europe. In the European Union, there are about 23 million micro-, small- and medium-sized enterprises, comprising 99% of all registered business subjects. The enterprises employ close to 75-million people and constitute the basis of the economic growth strategy (Medaković 2012a, b). The business of small- and medium-sized enterprises, i.e. the entrepreneurs, is burdened by numerous problems, but vary based on the phase of life cycle of an enterprise. At the start of the business, entrepreneurs are often adequately not prepared to launch their business and do not perceive all problems they will encounter. On the other hand, there is an evident lack of readiness of banks to finance the "start-up" activities of future entrepreneurs. In the phase of growth and development, small enterprises are burdened with other additional problems. As the first, management is burdened by achieving efficiencies to overcome a lack of information about the possibilities of growth and development, as well the challenges of market recognition and evaluations of their business' chances and risks. As the second, the transition from the entrepreneurial to the management phase of an enterprise causes a crisis in the form of lack of management skills, which is manifested in the lack of, or inadequate, business plans, intuitive models of business decisions making, inadequate controlling methods, lack of education, or criteria and plans, in general. As the third, even successful small enterprises face a crisis due to the needs of further growth, and that requires greater technical and administrative knowledge related to export business, international marketing and financial operations, professional knowledge on modern production and quality systems. As the fourth, with the growth of an enterprise and the increase of the technical-technological level of the business, there occurs a lack of professional labour force, modern production equipment, adequate business premises/objects and/or a suitable location. As the fifth, they must meet the challenge of financing growth through self-financing, credits for working capital and long-term investments, often with unfavourable conditions from perspectives of interest rates and mortgages.

The process of quick technological changes, which characterizes the modern economy, is mostly the result of new entrepreneurial ideas (West and Meyer 1997). Because of that, entrepreneurship becomes an opportunity for the countries in transition, because entrepreneurs can contribute to a greater degree of utilization of available resources and faster economic growth, by which it is possible to gain parity with developed countries (GEM 2012). As was shown by Ellis and Williams Max (2011), young entrepreneurs are confrinted by limitations to a greater extent, above all as regards access to finances and the creditability of the young, and at the same time such limitations decrease the chances for the young people to build their own business by themselves.

The development of small- and medium-sized enterprises is not realized only through the entrepreneurial practice and economic policy measures, but through system efforts to achieve direct, structural changes in the sphere of education. On the basis of the research up to now (Ćosić 2007), an increasing number of young people will link their professional development with the sector of small and medium-sized enterprises in the future. Many learned lessons and experiences (Dollinger 2005), extracted from successful projects, confirm that entrepreneurs should have the following potentials and abilities: a sharp perspicacity for new market chances and consumers' needs; general and personal management skills, good communication with potential investors/financers, partners, employees; analytical dexterity, enthusiasm, self-confidence and innovativeness.

Due to that, education must meet the requirements for new knowledge and abilities necessary for successful work in this sector, and that also implies the introduction of new occupations and profiles that will enable the achievement of a greater flexibility in dealing with the variable requirements of work.

2 Defining Entrepreneurship

In theory and practice, we often meet with the dilemma of making distinctions between the functions of an entrepreneur and a manager. The difference exists, certainly, but the distinctions depends on the amount of entrepreneurship actually used in particular enterprises, i.e. on the way that an enterprise is run. In every case, it is impossible to imagine the business of an enterprise outside the normal chain (ownership—market—entrepreneurship—management). So, it is impossible to imagine the function of a manager without entrepreneurship, and, at the same time, it is impossible to imagine an entrepreneur who does not use at least some of managerial methods.

The history of entrepreneurship dates back to the Middle Ages, and entrepreneurship came into full bloom with the development of capitalism, which unites the preferences for personal wealth and the wish for profit maximization. Over time, entrepreneurship and entrepreneurial activities attained a professional and civilized character. The notions of entrepreneur and entrepreneurship were used for the first time by Richard Cantillon, an English economist at the end of the 17th and the beginning of the 18th century, and, until today, an entrepreneur is considered to be an initiator and organizer of jobs, an innovator, a vigorous business man who operates under conditions of, at least, moderate risk and under its own property liability. From this statement, it can be concluded what is the strength of the entrepreneurship spirit and why that spirit is what forms the heart of the spirit of a market and the economic system in general. Zombart thought that the spirit, form-order and substance were the three key formative elements of every economic system.

Entrepreneurship is often defined even more broadly (Ellis and Williams Max 2011), and it is related to the type of behaviour, and not only the running of a business, or self-employment. For example, the International Labour Organisation (ILO) defines entrepreneurship as the way of thinking and reasoning. That is much more than the bare running of a business. That is a process in which individuals become aware of the options for self-employment and career building, development of ideas, taking and management of risks, learning the process of taking initiative in the development and owning of a business (Schoof 2006).

The theorists in the area of entrepreneurship are obsessed with defining the word "entrepreneur." Their priorities are maybe backwards. The entrepreneurship theory, as stated by Bull and Willard (1993), can solve the problem with a definition, and it becomes irrelevant.

Thus, for example, a famous theorist (Bygrave 1997) pointed out the nine primary factors of entrepreneurship. Due to the fact that all the notions (in English) start with "f", the concept has been known in literature as the 9F concept of successful entrepreneurship:

- Founders: significance of first-class founders, wise and respected;
- Focused: directing towards selected market segments (so called niches);
- Fast: fast decision making and fast application of the decisions;
- Flexible: fast adaptation to change;
- Forever-innovating: permanent quest for the new;
- Flat: permanent struggle to decrease all costs;
- Friendship: to be a friend with all stakeholders;
- Fun: it is fun to cooperate with an entrepreneurial enterprise.

It is clearly understandable that it is not easy to provide all the named factors simultaneously, as the assumptions of successful entrepreneurship, but it is undoubtedly necessary to strive to create a greater number of them, thus providing successful conditions for a good start of a business venture. Although it has been believed that a good start is what the entire entrepreneurial process depends on, a weak start will not disturb a true entrepreneur and make him/her quit. On the contrary, there is a not so small number of those, who draw additional inspiration and new energy, so that they persist and end up as winners. So, for a success, a synergy of the personal characteristics of the entrepreneur and business-environmental conditions is necessary.

Although it is not disputable that, for the manifestation of entrepreneurial behaviour, besides the personal features of an entrepreneur, being an individual and the holder of personality, it is extremely important that there are adequate socioeconomic, legal and cultural ambients in which personal features of potential entrepreneurs can be realized and developed. At first, it is necessary to check if and to what extent there are participants who have psychological and other potentials to become entrepreneurs in the given environment. The characteristics of potential entrepreneurs are the following ones (Medaković 2012a, b):

1. Taking the risk. An entrepreneur is a person who takes risks, i.e. someone who deals with a new business run outside pre-existing organizations.
2. Independence. An entrepreneur is an independent person who does not like to work for others, but who rather works on his/her own.

3. Internal centre of control. Those who believe that they themselves have a great influence on what happens to them are denoted as people who have an internal centre of control.
4. Live though excitement. Although some people strive for a calm life, some other are oriented towards excitement and action. Entrepreneurs fit into the second category, because they like the excitement brought by the new and different.
5. Self-initiative. Entrepreneurs start things when they decide to do so, using their own ideas and energy as the rationale.
6. Self-confidence. Entrepreneurs have great self-confidence, i.e. significant belief that the new product, service, idea or approach they propose brings benefits. This self-confidence is necessary because others often do not support their new ideas.
7. Adaptability. Entrepreneurs are adaptable. Markets and systems change. Business must change together with them.
8. Pertinacity. Entrepreneurs are pertinacious. They are the people who do not quit in spite of obstacles.
9. Ambiguity. Entrepreneurs must tolerate the dilemma of choice placed in front of them. For some persons, some situations are difficult to be resolve, dangerous and psychologically problematic. That is not the case when an entrepreneur is in question, because the uncertainty affects them less, so they are even able to use it.
10. Identification patterns. Entrepreneurs are capable of solving disputes by using identification patterns. They are able to see how everything fits together, perceiving in one glance the entire "wood," while others choose to fight with "trees".
11. Little need for assistance. This is closely related to self-confidence. Entrepreneurs are more than self-confident. They feel little need for assistance from other people.
12. "The right thing". The phrase presumes a psychological perspective in entrepreneurship. It is necessary to possess certain features, i.e. the characteristics of someone who wants to be an entrepreneur.

Besides the listed items, it is also interesting to mention the "10D" concept (Bygrave 1997), which identifies ten characteristics of a successful entrepreneur, simplifying the description of an entrepreneur, so it may seem to us that it is exactly the entrepreneur who we know from our neighbourhood. He points out the following ten features of a successful entrepreneur:

- He/she is prone to dream, because he/she has a precise vision of a certain business in the future, and, what is more important, to turn dreams into reality;
- He/she exhibits decisiveness, because he/she makes the decisions quickly and that is the key to his/her success;
- He/she is a doer, because what he/she makes, he/she realizes very fast in practice;

- He/she exhibits determination, because he/she rarely gives up even when he/she faces the most difficult problem and is determined to persist to the end in the realization of the idea in which he/she strongly believes;
- He/she exhibits devotion to business and friendship relationships (business partners) and is ready to help escpeically when the situation is the most difficult;
- He/she pays attention to details, because he/she does not allow himself/herself to miss even the smallest thing;
- He/she is the master of his/her destiny, because he/she holds everything in his/her hands;
- Money (dollars) is the measure of his/her success, but he/she does not have the aim to accumulate wealth, rather he/she sees money only as the means of expanding a business venture;
- He/she is ready to assign tasks, because he/she is ready to delegate the authorization to key associates, even, before that, to members of his/her family.

Which of these characteristics—they can also be called the entrepreneurial values—will be accepted, rejected or ignored in the environment in which an entrepreneur operates, depends on numerous circumstances, and above all on accepted social norms, as well as on economic, political, and, increasingly, on international conditions. Politically and economically stable societies are more tolerant towards the stated values, because they are widely accepted and make up a part of the social and entrepreneurial climate, culture and moral.

Besides the analysed personality characteristics of an entrepreneur and the structure of knowledge he/she must possess, it is necessary to say something on the specific skills he/she needs to perform such a complex and multi-layered activity as entrepreneurship.

Above all, it is necessary to remember that his/her entire work is reduced to the realization of an idea to achieve a positive result and that is always together with other people, never alone. An entrepreneur needs skills so that he/she converts the ideas into reality. The conversion of ideas into reality requires two types of skills from an entrepreneur: management ability in the purpose of organizing physical and financial resources and human resources management skills because of the necessary support of others. Every entrepreneur, regardless of his/her abilities and education, has to rely on associates, and that is why teamwork skills and abilities are so essential for him/her. It is also very important that an entrepreneur possesses the following skills (Medaković 2012a, b):

- strategic ability, i.e. the ability to perceive a job as a whole, holistically, but also the ability to look ahead, prospectively;
- ability to plan as an initial step every action and ad a basis of perceiving the future, but also the ability to predict the future;
- marketing ability, by which the needs, demands and interests of clients as the key stakeholders are perceived, because of which, generally, an entrepreneurial venture has been initiated;

- financial abilities, because the management of money as working capital is the condition that ensures the continuity of a business and the flow of money, and the ability to evaluate the required investments and accompanying risks;
- ability for project management, which implies project organization and providing project resources at the right time and at the right place;
- ability for time management, since time is a resource that cannot be refunded, implying a strict hierarchy of priorities in performing the tasks.

Besides the named abilities and skills of an entrepreneur, it is useful to indicate the necessity of possessing the abilities which qualify him/her as a successful communicator within an enterprise's environment. Above all, in this regard, we think of (Medaković 2012a, b):

- management skills, needed to stimulate the people to work and perform the tasks important for the success of an entrepreneurial venture, their directing, support and assistance in realization of the tasks, organization of work processes, selection of staff and rational use of their abilities, administration and control of all work processes and achieved business results;
- motivation ability, by which the commitment to an entrepreneurial idea, jobs and tasks in its realization are achieved; it is also important that an entrepreneur equally motivates himself/herself, and also the people he administers in the business process;
- ability for the distribution or assignment of jobs and tasks, assuming knowledge about the abilities the people possess and the means of their engagement and of greater development of their abilities for a more efficient execution of jobs;
- communication abilities, i.e. the need for oral and spoken expression of ideas and informing of people about them with the aim to affect the actions of people by means of information and communication;
- negotiation abilities are not bargaining abilities, but the creation of a situation of general understanding and recognition of possible outcome for all the sides in a negotiation process;
- operational ability, as a response to the changes in a dynamic organisation system and as a condition for a successful and efficient business by taking adequate measures and actions;
- creativity, as the ability for logical thinking and inference, which produces the ideas and practical solution to problems by successful use of markets, technical aspects, staff, scientific details, other information and other potentials;
- ability for critical observation of statuses and relationships because nothing should be taken "for granted," but every bit of business information and phenomenon must be critically discussed, its positive and negative effects perceived, and only after that satisfactory decisions should be made:
- analytical ability enables a complex organizational system, composed of numerous elements and subsystems, to be perceived, arranged and functionally harmonised with numerous internal and external factors by entrepreneurs.

The authors (Zimmerer and Scarborough 2005) also allude to these facts, stating the following characteristics of entrepreneurs:

1. Wish for responsibility. Entrepreneurs feel a deep responsibility for the ventures they have started. He/she wants to control their own resources and use them to achieve the set goals.
2. Readiness to take a moderate risk. Entrepreneurs are not too comfortable with taking a risk, but are ready for a moderate, calculated risk.
3. Trust in their own ability to success. Entrepreneurs usually have great trust in their own ability to succeed and are optimists in terms of their chances of success.
4. Wish for a quick reaction. Entrepreneurs enjoy running their businesses and want it to be known that they make business decisions quickly and independently.
5. High level of energy. Entrepreneurs are much more energetic then ordinary people. The energy can be a critical factor in starting own enterprises.
6. Orientation towards future. Entrepreneurs have a well-defined feeling for finding opportunities. They look ahead and are less burdened with what they did yesterday, taking more care about what they will do tomorrow.
7. Skill in organization. Entrepreneurs know how to choose specific people to perform a specific task. By efficient combination of people and jobs, they achieve the transformation of their own visions into reality.
8. The wish for success before money. It is wrong judgement that entrepreneurs are motivated only by the wish to gain money. On the contrary, the wish for success is in the first place for entrepreneurs, and the money is only one way of measuring the achievements.

3 Profiles and Characteristics of Entrepreneurs in the Balkans (B&H)

The current globalization process convinces use every day that the so-called global entrepreneurial revolution is in progress in which the main participants become the economies of China and India (not so significant until recently), and of many other Asian countries, as well as of Russia and Brazil. The abundant talent for creation of almost incredible business ideas, a flood of the most diverse production and service programs and projects, unbeatable talent for innovation, and especially for imitation, great inner markets and aggressive orientation towards exports are all sufficient reason to reassess the up-to-now trajectories of entrepreneurship based on the Western paradigm of selfish individualism and quick enrichment. In the context of such global transition processes, the profile of an entrepreneur as a key agent of change, who confronts great challenges, should be redefined, i.e. redesigned. At the beginning of the 1970s, the so called third technological (microelectronic) revolution started in the West, which really rocked the world of labour thoroughly, especially that of industrial production, and it contributed to sophistication and systematization of many spheres of human life. In the 1990s, completely new opportunities, the so-called e-business and net entrepreneurship, were launched along with revolutionary breakthroughs in trade, banking, the stock-exchange business etc. But, relatively quickly, the fast-growing business (dot.com) experienced a deadlock, and, at the turn to this century, even a collapse, with dramatic consequences. It is recovering at the moment and will probably gain new strength, but a logical question occurs on the real perspectives of new businesses, their reach, limits and likely prospects.

In that light, new roles, tasks and, especially, the profile of modern and future entrepreneurs are being studied. Essentially, the roles and tasks of entrepreneurs will remain more-or-less the same, including the fact that the way to realization of the roles and tasks will experience many changes. It is apparent even now that the main change is reflected in the change of the entrepreneurs' focus from financial, material, and, especially, technical-technological resources, to human and intellectual resources as the key components of every entrepreneurial process.

Some important characteristics of a modern entrepreneur then are (Medaković 2012a, b):

- A modern entrepreneur is not a rigid boss, slave driver who is awe-inspiring, but is also a leader and trainer who commands respect, because he/she knows how and wants to encourage, motivate and develop the creative energy of the members of his/her team in the realization of common goals.
- A modern entrepreneur is dominantly oriented towards the stimulation of creativity and innovativeness. He/she does not wait for changes to react, but generates or at least anticipates them, i.e. predicts and prepares himself/herself for upcoming changes. He/she is not afraid of change, because they are seen as a challenge and a chance for new success.
- A modern entrepreneur raises the morale and works on the improvement of business ethics in his/her environment. He/she is aware of the long-term importance of a business' reputation, and business ethics can contribute to that in many ways.
- Because of the strengthening of his/her own abilities, as well as the abilities of the associates, a modern entrepreneur becomes aware of the importance of constant renewal of knowledge and skills necessary for following, predicting and generating changes.
- The ability for effective communication becomes an imperative for success for modern entrepreneurs. This fact must be recognised in all educational curricula for entrepreneurship, and, above all, in business schools.
- A modern entrepreneur is greatly oriented towards the global business stage. Because of that, he/she must be familiar with international relations, relations among countries and people, various cultures and economies and must be capable of developing the goals and strategies of his/her enterprise in the wider context of international business and fiercer competition.

- Modern entrepreneurship becomes more and more based on high technology, especially information technology. Due to that, a modern entrepreneur has to know the key trends (megatrends) of production and control technologies and to take them into account as the key factors for gaining a competitive advantage.
- Modern entrepreneurship is based on legally regulation and the procedures that are internationally standardised and harmonised to a great extent. Because of that, a modern entrepreneur must know the logic of complexity in the field of legislation and respect the legality of competent local and international institutions and accepted conventions, treaties, agreements etc.

In brief, modern entrepreneurs are a personification of numerous human qualities and features, and because of that they can be classified into these three groups (Medaković 2012a, b):

- Entrepreneurs–technicians, when there is a wish to point out their emphasized technical orientation, which, above all, considers them as innovators in the creation of new products or processes. In that, they see the creation of organisation as the instrument for realization of their defined goals, and not as the final goal.
- Entrepreneurs–creators of organisation, to whom the creation of organisation is the most important goal, as a opposed to entrepreneurs–technicians. The most important thing for them is to realize the development of organisation by means of the growth and development of business, which is measured by the increase of profit and the number of employees. They are aware that it is possible, through good cooperation with people, by organising them into teams within which they exert their influence and power.
- Entrepreneurs–job constructors, who are the "cause" of the processes of contracting, negotiating and making new arrangements, because they like initiating jobs. They do not tolerate a total dedication to one organisation over a long period of time and are always ready to start a new entrepreneurial venture because they enjoy the creation of a new business and always have a "backup option."

Although the territory of the Balkans is incorporated into the processes of European integration and global processes, there are still numerous specificities of doing business there, especially in so-called transitional countries. Such specificities are reflected, above all, in the long-term isolation and enslavement of some nations, in the major influence of tradition, in the mixture of cultures, religions (exchanges, of Eastern and Western Christianity and Islam), etc. They have a manifold origin and are not only resident, but are also formed in modern conditions that are marked by numerous crossed interests of a wide circle of (international) participants at the "geostrategic route" of Europe and which, in most cases, to not have legitimacy (justification) for their actions. Having all that in mind, it is also possible to speak about a specific type of a so-called Balkan entrepreneur. However, in doing that, the significant differences between particular Balkan countries and societies must be taken into account, because some of them have always been capitalistic, while others are in the process of transition to capitalism, and some are formally integrated into European integrations, so that it gets more and more difficult to talk about a unique profile, although the differences in relation to the Western European profile are still very visible. It also have to be pointed out that there are many entrepreneurs who, in their features and behaviour, do not differ in any way from the best European entrepreneurs and managers. Unfortunately, they are still not a majority. In brief, we can state a few important flaws of Balkan entrepreneurs, which should be removed as soon as possible (Medaković 2012a, b):

1. The lack of vision and clear perspective on what to do in a turbulent business environment;
2. Wrong or completely neglected following of the flow of money in most enterprises;
3. Insufficient knowledge and understanding of the structure of costs;
4. Nonexistence of the practice of making business plans for an enterprise;
5. Underdevelopment of a marketing approach to entrepreneurship;
6. There are no traditions, skills and techniques for rapidly making important decisions;
7. There is very strong distrust towards all forms of business cooperation, especially with foreigners;
8. Ethical standards of entrepreneurship and personal fairness are not on such a high level;
9. The age structure of entrepreneurs is unfavourable, i.e. there are very few young entrepreneurs in the Balkans in ages up to 30; most of them are up to 50 or older;
10. Mental inertness and aversion to further education;
11. Persistent orientation to waiting for external help (there is a strong inclination by some Balkan entrepreneurs towards waiting for someone (i.e. the state or an international organization, e.g. the European Union funds, to determine finally "our development priorities", with every entrepreneur hoping that exactly his/her enterprise will be included in the priorities, which would solve all problems automatically);
12. The lack of understanding and slow acceptance of the ecological criteria of business.

3.1 Characteristic Factors Which Enable Entrepreneurs to Recognise the Chances for Business

Potential entrepreneurs, i.e. the ones who possess certain characteristics, can recognise the chances for running a business more easily, and the factors are:

Previous experience. A previous experience in a certain branch of the economy helps an entrepreneur to recognise business opportunities. Working in a certain branch of the economy, an individual may notice an underserved market niche. It is also possible that, working in a certain branch of economy, an individual may create a network of social contacts which provides information from the inside, and which leads to recognition of new opportunities. Once an entrepreneur starts an enterprise, new business opportunities become obvious.

Cognitive factors. Recognition of opportunities can often be an inborn feature or a cognitive process. Some people think that entrepreneurs have a "the sixth sense" that enables them to notice the opportunities other people fail to see. This "sixth sense" is called entrepreneurial consciousness, being formally defined as the ability to notice things without intentional research. Most entrepreneurs consider themselves being like this, believing that they are more capable and conscious in relation to other people when the recognition of opportunities is in question. For example, a computer-science engineer, in relation to a lawyer, will notice better the needs and opportunities in the computer industry.

Social networks. The scope and depth of a social network of an individual affect the recognition of an opportunity. The persons who create a significant network of social and professional contacts will be exposed to a greater number of opportunities and ideas than the persons with scarce networks. The connections with other people are known under the name "relations". We are all in certain relations. Strong relations are characterized by a strong structure and frequent interaction between the job colleagues, friends and spouses. Weak relations are characterized by a weak structure and a not-so-frequent interaction between accidental and superficial acquaintances. As an example, take an electrician who explains to a restaurant owner how he solved a certain problem. After he has listened to the electrician, the restaurant owner may say: "I would never hear about such a problem solution from someone from my enterprise or branch of economy. Such an approach to problem solving is completely new to me, and it could exactly help me to solve my problem".

Creativity is the process of creating a new or useful idea. Recognition of an opportunity can, at least partly, be seen as a creative process. It is easy to notice creativity incorporated into the creation of many products, services and businesses. In an increasing number of enterprises, the source of creativity are the employed entrepreneurs.

Figure 1 shows the process of recognizing the opportunities for running a business. As it is shown there, there is a connection between the awareness of variable trends in the environment and the personal characteristics of an entrepreneur, because these two aspects of opportunities are mutually dependent. For example, an entrepreneur with an established social network can be in a better situation to recognise the emerging technological trends in relation to an entrepreneur with a modest social network. Or, the awareness on the emerging technological trends, such as digitalisation, may support entrepreneurs to participate in conferences and workshops on the topic, thus propagating their social networks.

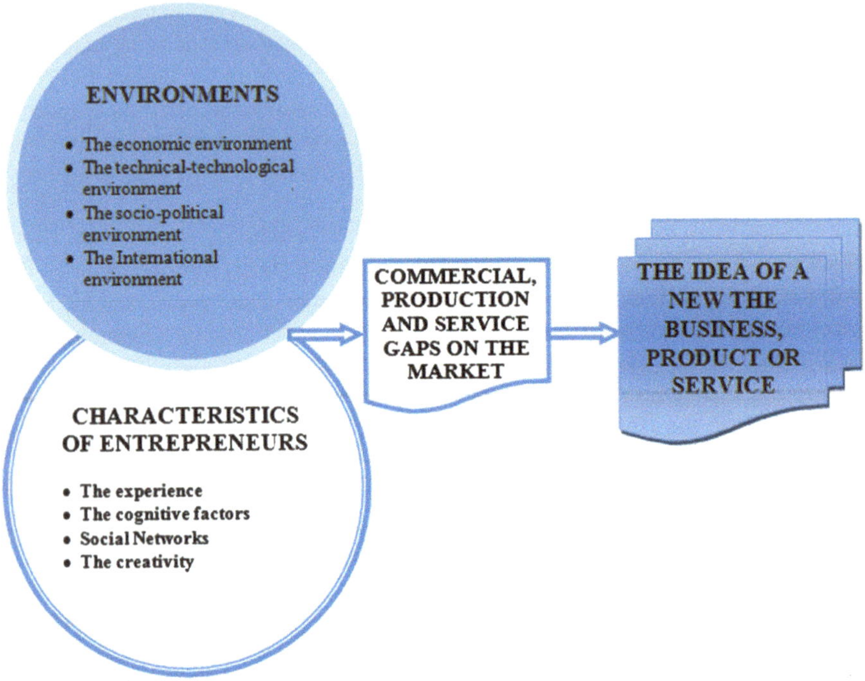

Fig. 1 Process of recognising the opportunities for starting a business

4 Motives of Entrepreneurs

The motives of various entrepreneurs are not the same (OECD 2004). Young entrepreneurs have different needs and, by means of that, different motivation for running a certain business. A few studies presents diverse ways of classifying young entrepreneurs (Ellis and Williams Max 2011).

However, as stated by Desai (2009), entrepreneurs take the opportunity informally and only formalise it after they have realized significant benefit from it. This is mostly the case in the developing countries where local markets propagate very fast, thus making the opportunities for a new beginning and business, of course, with a favourable political environment and arranged legislation, because otherwise a deadlock in economic growth and development may occur.

A motive can also be an aspiration for potential growth and ambition towards international markets, as well as the level of innovations in products and services offered by entrepreneurs in an entrepreneurial process, with the aim to create new values and to open new job positions (Ellis and Williams Max 2011). Young entrepreneurs have a tendency to have a greater potential of growth than the older entrepreneurs, i.e. the entrepreneurs of an older age.

Empirical research on the perspective of entrepreneurship and entrepreneurs, from the aspect of motivation and behaviour, that may lead to creation of new job positions, i.e. enterprises, as well as enterprise theories, in most papers, rely on the process of perceiving and analysing without inclusion of the enterprises themselves in the process (Barba-Sanchez and Atienza-Sahuquillo 2012).

There are various motivation theories that try to explain the behaviour of the employed as a whole, while very few of these can be applied to the theory when studying entrepreneurship and entrepreneurs.

In the paper of Barba-Sanchez and Atienza-Sahuquillo (2012) are shown the following motives which lead to an entrepreneurial venture, and these are, above all: the wish for independence, affinity for risk and affinity for innovation. The decision to start a business, i.e. a start-up, should be observed as the product of motive and intention, and the entrepreneurial features and abilities should be differentiated on the basis of that.

Various entrepreneurial motives follow different intensities of the factors (Barba-Sanchez and Atienza-Sahuquillo 2012). On the other hand, the motivation system of a certain person should have a decisive influence to his/her behaviour, but that is not the only determinant. The perception and subjective opinion of a person depend on his/her internal nature. Inborn abilities of certain persons, as well as potentials and variables, such as effort, ability, previous experience, age, education, family heritage and environment, affect the behaviour and motivation of a person to decide to become an entrepreneur and establish his/her own business (Barba-Sanchez and Atienza-Sahuquillo 2012). It is also necessary, to take into account some other facts, such as the need for affiliation, need to achieve power, tolerance, innovations, readiness to take a risk and persistence.

What usually initiates a person to start a business is achieving her/his own pleasure, and also the will to power, but the other thing is generally less important. The need for independence, as well as creativity, are always part of the basis of the development of one's own business project, i.e. a business.

Motivation factors include: readiness to take on a challenge; pleasure in work and solving difficult projects; setting personal goals; talent, development of potentials and progress to achieve efficient results; control of resources; pleasure in work with experts; and constant striving for performance improvement.

The other motivation factors include wishes, such as: to be an own boss without the need to be responsible to superiors; pleasure in running a business, as well as in guiding other persons pursuant to the predicted plans; mobilization of resources and engagement of labour, the people, to accomplish their own goals; taking on responsibility; pleasure in competition and defeating the competitors; identification of personal abilities and to be the centre of attention; and creation of reputation and status in the society.

Some other factors include: wish for independence from other people and limitations; removal of limitations and directives; wish to evaluate one's own decisions; unwillingness to depend on other people on the emotional or social plan; insistence on freedom; wish to assume changed behaviour; pleasure in presenting and giving form to creative ideas; and the urge to produce new products.

5 Empirical Research into the Preferences of Students of Entrepreneurship

The paper (Medaković et al. 2016) shows the analysis of the acquired competencies and the requirements for starting a business, i.e. discussing their personal efficiency, organisation and motivation. Research by Maresch et al. (2016) has shown that education in entrepreneurship increases entrepreneurial intentions, however, this does not happening uniformly in all contexts, and suggests that future research should be directed toward an effective didactic approach for students of engineering sciences. The study of Van Gelderen et al. (2015) is based on the Rubicon model of action phases of learning, action or lack of action, which follows the establishment of entrepreneurial intentions. At the same time, they examine the role of self-control and emotions associated with future intents. They showed that positive self-control alleviates the relationship between the intentions and actions of entrepreneurs. The work of Shinnar et al. (2014) shows the role of entrepreneurship education in the strengthening of entrepreneurial efficiency, by providing business training and, specifically, training to prepare students for business planning. Premand et al. (2016) point to the potential that enables young people to acquire skills and start their own business. Paula and Shrivatava (2016) focused on the entrepreneurial intention in the context of conducting stronger entrepreneurial intentions, comparing young managers in developed countries and young managers in non-developed countries. They have analysed the link between entrepreneurial intentions, the country's culture and proactive behavior.

5.1 *The Aim, Methods of Research and Description of the Sample*

The aim of the research is to determine the levels of awareness of students studying engineering disciplines in terms of entrepreneurship, as well as certain attitudes and competencies related to entrepreneurship. The study applied statistical methods, empirical research, methods of logical explanation and comparative analysis. The method of sampling was used in the analysis and the selection of an appropriate sample for research. The research was conducted on the survey list, which is made up of two separate tests. Test 1 relates to the fulfillment of conditions for business and consists of 22 questions of multiple-choice form (a, b and c) which are valued differently, and test 2 relates to the tendency toward entrepreneurship, consisting of 25 statements that were confirmed or refuted by spanning five levels, from completely agree to completely disagree. Empirical research was conducted at the University of East Sarajevo, i.e. among students studying an engineering discipline. The conclusions are derived by the methods of induction and deduction.

The sample was performed with 145 students, i.e. students of engineering disciplines, all years of study, which may be representative and fully respond to the requirements set out for the investigation.

5.2 Results and Discussion

Processing the completed questionnaires gives the following results, which are summarised presented in Figs. 2 and 3. Figure 2 presents the fulfillment of conditions for business.

In Fig. 2, we can see a very small percentage of respondents, a total of 3%, which possess a suitable personality for being entrepreneurs, which indicates a very weak entrepreneurial awareness among the population of students of engineering disciplines. From the aspect of research, we came to the fact that 49% of respondents possess certain characteristics of entrepreneurs, but would require consultation with experts, and from that point there is enough potential. A course in entrepreneurial engineering could be improve the situation to some extent and activate the potential of entrepreneurial characteristics. Thus, in the future, they will be willing to start their own business in the area of their education, and beyond.

From a total of 145 students surveyed, test 1 under a) and b) rounded according to the following key:

- 36.8% of the surveyed students gave answers relating to the first seven questions, which suggests that there is a problem of psychological abilities, concerning which in adults it is difficult to influence.

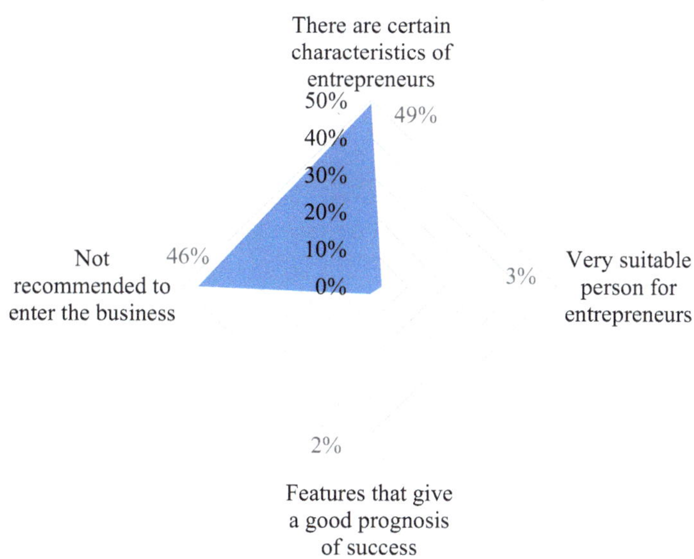

Fig. 2 Fulfillment of conditions for business

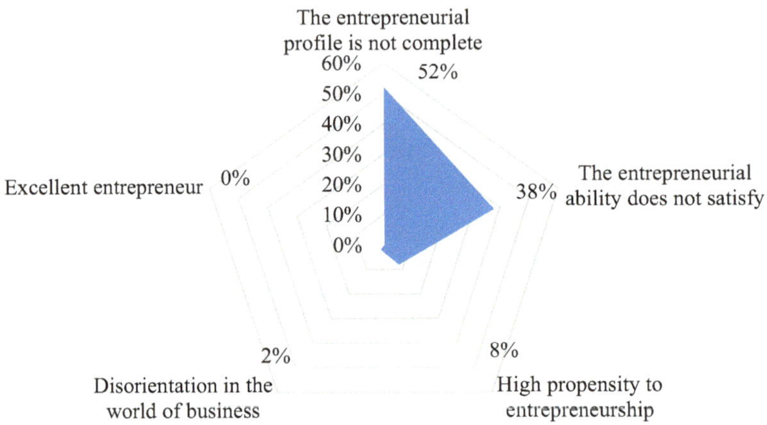

Fig. 3 The tendency of students to entrepreneurship

- Answers to questions 8—10 relate to health problems, eyesight and hearing, which can eliminate certain adjuvants; otherwise you can expect interference in business, and this constitutes 10.76% of the total number of students surveyed.
- 11.37% of surveyed students gave answers to questions 11–13, which relate to personality, and about which the entrepreneur must take care; this recommends a detailed study of the subject matter of personality traits.
- 41.03% surveyed students gave answers from questions 14–22, related to knowledge and experience in business, as well as the attitudes and opinions related to the essential characteristics of a successful business.

Figure 3 presents the results of test 2, which referred to the tendency towards entrepreneurship.

Of the total surveyed students, no student showed traits of an excellent entrepreneur. A high tendency to entrepreneurship showed up in 8% of respondents, which is also devastating, while 52% of respondents did not meet the entrepreneurial profile, and they need the advice and help of others.

On the other hand, 38% of the surveyed students did not meet entrepreneurial skills, and 2% of respondents indicated no orientation in the world of business.

6 Conclusion

The definition of entrepreneurship has evolved over time. The earlier understanding of entrepreneurship is not similar to the current one and will acquire yet other attributes in the future. Basically, there is a consensus in some aspects of definition, such as: creativity, innovativeness, independence, and affinity for risk. Entrepreneurship does not come to the fore outside a market ambience. Entrepreneurship and its creative components are related for the idea of market verification, i.e. the commercialisation of a result in the market. The main characteristic of entrepreneurship is the creation of something new in a production or service program, depending on influences from the environment.

It can be said generally that a delay in the development of small and medium-sized enterprises, as well as of entrepreneurial awareness, i.e. the entrepreneurial activities in B&H, dates from the period of the 20th century. We must take into account that the economic structure of the country in the second half of the 20th century was based on large companies, which relied on the socialistic model of economics, a planned system, which did not recognize private ownership, business and free entrepreneurship, except some crafts which managed to survive. In that context, not only was the sector of small- and medium-sized enterprises, i.e. of entrepreneurship and private ownership, unsupported, but it was even restricted.

Entrepreneurship has a key role in the process of transition, especially in terms of self-employment. A great support from all areas of human activity is necessary, especially by educational institutions, to improve the entrepreneurial skills of the young.

The characteristics and profiles of entrepreneurs presented in the paper do not characterize, to the same degree, all those who consider themselves entrepreneurs. These characteristics comprise what could be called the typical entrepreneurial profile. Due to the fact that every person has her/his own special personal features, they make an entrepreneur unique. Of course, the number of features varies and is more expressed in some entrepreneurs then in some other.

If the entrepreneurial profile of a person is similar to the profile of entrepreneurs, in general, these persons can be expected to exhibit entrepreneurial behaviour and to be more strongly motived to start their own businesses. If certain conditions and events are favourable, the individuals with adequate entrepreneurial characteristics apply their resources and compensate for their weakness to exploit a sufficiently favourable situation. None of the discussed characteristics and profiles of entrepreneurs gives an integral interpretation of the key factors for success of an entrepreneur, but they integrally enable the understanding and explanation of the reasons for success of entrepreneurial enterprises and, of course, entrepreneurs as owners of such enterprises. Success is the result of the personal characteristics of an entrepreneur and the ability to learn from experience and to adapt to influences from the environment. The personal characteristics of an entrepreneur are the most important in the start-up stage of a new business. The ability to learn from experience becomes more important after the running of an enterprise, and the growth and development of the enterprise depend on the gained experience, knowledge and ability to adapt to influences from the environment.

Entrepreneurial intentions depend on the way in which an entrepreneur experiences the feasibility of the imagined idea and his/her affinity for action. The spark to action often depends on the availability of resources. Before starting a business, an entrepreneur usually has a form of a business plan. That is the entrepreneur's personal concept, vision and expectation about the way in which the business activity will develop. From that point of view, a business plan is desirable and relevant.

An entrepreneur may plan only the project, i.e. to design a business plan, but he/she usually depends on some assistance in the form of training, advice or consulting. His/her decision to take action in a certain moment may depend on whether he/she feels relevant and efficient.

Physical resources include technologies, equipment, processes, buildings and information necessary for work of business subjects. Finally, financial resources are often the key elements. Problems may occur both because of their unavailability and of their availability. An entrepreneur's ability must be reflected through resource management, and thus satisfy the needs and solve the problem. Furthermore, he/she has to believe that the resources exist and that they are available. The efficient, motivated employees have to be mobilised at all stages of business development, especially at running a business. In that period, an entrepreneur has to perceive everything and, at the same time, to enable certain employees to learn the new jobs. The ability to build competitiveness is of crucial importance for the success of an enterprise. Entrepreneurial behaviour depends on motivation, ability, and a person's attitudes and intentions.

The future of entrepreneurship is promising. The support to entrepreneurship, i.e. to entrepreneurs, is growing. There is even a global process that unifies, to a certain degree, the relationship of the state to entrepreneurship and the development of small- and medium-sized enterprises pursuant to that. It is apparent even now that the main change is reflected in the change of the entrepreneurs' focus from financial, material, and, especially, technical–technological resources to human and intellectual resources as the key components of every entrepreneurial process. From this point of view, educational institutions (primary and secondary schools, universities, etc.) must pay more attention to promoting entrepreneurial behavior, especially in developing countries. Looking at the overall situation, as well as the results of the research, we can conclude that there is some potential and that it should be properly directed to raise awareness and create entrepreneurial spirit.

References

Barba-Sanchez V, Atienza-Sahuquillo C (2012) Entrepreneurial behavior: impact of motivation factors on decision to create a new venture. Investig Eur Dir Econ Empresa 18:132–138

Bull I, Willard GE (1993) Towards a theory of entrepreneurship. J Bus Ventur 8:183–195

Bygrave WD (1997) The portable MBA in entrepreneurship. Wiley, New York

Ćosić I (2007) The possibilities and limitations of entrepreneurship development in selected municipalities in the province of Vojvodina. Project Documentation, Novi Sad 2007

Desai S (2009) Measuring entrepreneurship in developing countries. UNU-Wider. Research paper no. 2009/10

Dollinger M (2005) Entrepreneurship strategies and resources. Prentice Hall, Harlow

Ellis K, Williams Max C (2011) Imising impact of youth entrepreneurship support in different contexts—background report, framework and toolkit for consultation. Overseas Development Institute, London

Global Entrepreneurship Monitor—GEM (2012) 2011 Global Report. Edited by D. J. Kelley, S. Singer & M. D. Herrington
Kalyani B, Kumar MD (2011) Motivational factors, entrepreneurship and education: Study with reference to women in SMEs. Far East J Psychol Bus 3(3):14–35
Maresch D, Harms R, Kailer N, Wimmer-Wurmc B (2016) The impact of entrepreneurship education on the entrepreneurial intention of students in science and engineering versus business studies university programs. Technol Forecast Soc Chang 104(2016):172–179
Medaković V (2012a) Research of conditions for development of small and micro enterprises and development of entrepreneurial infrastructure model. Doctoral thesis. Faculty of Mechanical Engineering, East Sarajevo
Medaković V (2012b) Ability of managers for crisis management in SME's. In: II international conference industrial engineering and environmental protection 2012 (IIZS 2012). Zrenjanin, pp 351–356
Medaković V, Moljević S, Gojković R, Vasković S, Miljanović M (2016) Tendency to entrepreneurship of students engineering disciplines. (Serbian language Sklonost preduzetništvu studenata inženjerskih disciplina). J Qual Excell 1–2/2016:40–43
Melović B, Grubić-Nešić L, Mitrović S, Gashi Marko LJ (2011) Preconditions development of modern entrepreneurial society. (Serbian language Preduslovi razvoja modernog preduzetničkog društva). Entrep Mag 6:13–18
OECD (2004) Fostering entrepreneurship and firm creation as a driver of growth in a global Economy. Paper for the second OECD conference on "Promoting entrepreneurship and innovative SMEs in a global economy: towards a more responsible and inclusive globalisation". Istanbul, Turkey
Paula J, Shrivatava A (2016) Do young managers in a developing country have stronger entrepreneurial intentions? Theory and debate. Int Bus Rev. doi:10.1016/j.ibusrev.2016.03.003
Premand P, Brodmann S, Almeida R, Grun R, Baroun M (2016) Entrepreneurship education and entry into self-employment among university graduates. World Dev 77:311–327
Schoof U (2006) Stimulating Youth Entrepreneurship: barriers and incentives to enterprise start-ups by young people. SEED working paper no 76
Shinnar SR, Hsu KD, Powell CB (2014) Self-efficacy, entrepreneurial intentions, and gender: assessing the impact of entrepreneurship education longitudinally. Int J Manag Educ 12(2014): 561–570
Van Gelderen M, Kautonenb T, Fink M (2015) From entrepreneurial intentions to actions: self-control and action-related doubt, fear, and aversion. J Bus Ventur 30:655–673
West P III, Meyer D (1997) Temporal dimensions of opportunistic change in technology-based ventures. Entrep Theory Pract 22–2:31–52
Zimmerer TW, Scarborough NM (2005) Essentials of entrepreneurship and small business management. Prentice Hall, New York, p 5

Categories and General Characteristics of Entrepreneurial Infrastructure

Vladо Medaković and Srđan Vasković

Abstract The paper presents the general characteristics of entrepreneurial infrastructure. One of the modern ways to support small newly established enterprises, which are in a developmental life phase, and entrepreneurs is the system of technological infrastructure comprising: entrepreneurial incubators, technology centres, science parks and business zones. Those are the different organizations that help entrepreneurs to develop their business ideas and to overcome more easily the initial problems in business, for which, in a wider context, the term business incubators is used, and also the clusters related to entrepreneurs who are in an advanced phase of entrepreneurship. An incubator is extremely suitable for newly founded small enterprises, which do not have their own business premises, sufficient funds and experience, on one hand, but, on the other hand, they have entrepreneurial ideas, goals and determination to do business. The opportunity to give local and regional support to newly founded and small enterprises by means of business incubators in Bosnia and Herzegovina is significant because it delivers the key elements for the development of incubators, such as unused spaces in all municipalities, that can be easily transformed into a workspace and adapt to the needs of new entrepreneurs. Incubators can function independently or they can be, for example, a part of a science park. In this paper, we present in the tables existing organizational forms of business infrastructure (clusters and entrepreneurs—business centers: free zones, industrial and business zones, science and technology parks, business incubators), their activities, and numerical strength in the territory of the Republic of Srpska.

Keywords Entrepreneurial infrastructure · Business zones · Incubators · Cluster · Entrepreneurship

---

V. Medaković (✉) · S. Vasković
Faculty of Mechanical Engineering, University of East Sarajevo, Vuka Karadzica 30, East Sarajevo 71123, Republic of Srpska, Bosnia and Herzegovina
e-mail: vlado.medakovic@gmail.com

S. Vasković
e-mail: srdjan_vaskovic@yahoo.com

J. Peña Dopazo and F. Zivic (eds.), *Supporting University Ventures in Nanotechnology, Biomaterials and Magnetic Sensing Applications*,
DOI 10.1007/978-3-319-61237-9_2

1 Introduction

Infrastructure is important for entrepreneurial activities (Heger et al. 2011) and may have various forms and functions. As first, the development of trade and industrial growth require physical infrastructure, road and railway traffic and transportation etc.

After the World War II when the state had nationalized private property (Medaković 2012), the spirit of entrepreneurship in the territory of former SFR Yugoslavia, including B&H as well, was cut off at the roots. The economic philosophy of the new economic system was based on the criticism of all the aspects of capitalism, such as individualism, private ownership over the means of production, entrepreneurship in a wider sense etc. Entrepreneurship was identified with private ownership. In the beginning (Cvijić 2007), that had meant small business. However, with further breakthroughs and the propagation of small business, the term private entrepreneurship occurred, which, as such, was accepted in official frameworks. Specifically, in the economic structure of the SFR Yugoslavia, especially after 1976, once The Law on Associated Labour had been adopted, a significant number of labour organisations (enterprises) emerged, of a combined type, in the fields of mining, metallurgy, metal processing industry, military industry and the wood-processing industry, with a great number of basic organizations of associated labour and complex organizations of associated labour, which had disregard of market laws as a common feature. Such enterprises, being labour-intensive, were oriented to the employment of a great number of people. The society propagated the need of full employment. There was a general sense of security, especially once a state job had been obtained. No one thought about any significant individual engagement in the area of entrepreneurship.

The structure of the individual sector consisted of agriculture, then of independent forms of production, service and construction craftsmanship, and independent catering, independent car-transportation activity and independent trading activity. In the SFR Yugoslavia, the percentage of the employed, as stated by (Cvijić 2007), in small enterprises with 7–100 employees was only 2.4%, and, in enterprises with 1–6 employees, was 5.6%. The number of employees was limited to 10. In the process of building socialism, the private sector was called small business, and it operated under numerous limitations in terms of what it could deal with and with whom and how many employees it could engage.

Such conditions are a phenomenon which is characteristic for all former socialist countries, and which is related to a lack of small- and medium-sized enterprises in the economic structure, and, by means of that, to the absence of creating an entrepreneurial infrastructure.

In all developed Western countries and in many developing countries, entrepreneurship and small enterprises as a whole are supported by the state, state institutions and nongovernmental organizations in many ways (Medaković 2012).

Such an orientation of the market-developed countries has deep roots, regarding the fact that capitalism has tried many developmental models as opposed to the one-dimensional models of economic flows control that have been practiced more-or-less for decades in the countries of socialist and similar socio-economic systems.

Similar to the leading countries of the West, many small countries, which started with the implementation of market-capitalistic principles in the development of economy three to five decades ago, have reached an enviable level of development today (Medaković 2012)—exactly due to the development of small enterprises.

The determinations of Bosnia and Herzegovina (Cvijić 2007) related to the SMEs development sector rely on the recommendations of the European Charter and the Act on Small Business. The Law on Ministries and Other Control Bodies of Bosnia and Herzegovina has also defined the institutional framework in the field of issues in the sector of SMFs, whose difficulties reflect, above all (Cvijić 2007): approaches in defining policies, development strategies and goals in the sector of SMFs; competences; means of developing a harmonized monitoring of the results in this area; mutual cooperation and profitability; and excessive administration.

At the level of the Republic of Srpska, within the Ministry of Economy, Energy and Development, there is a department for small and medium-sized enterprises, the head of which is an assistant minister with the responsibilities in the labor field including: development of entrepreneurship and craftsmanship; making of medium- and long-term development plans; and forming the developmental strategies for SMEs and entrepreneurial activity.

Pursuant to The Law on Enterprises of the RS (2006), an enterprise is a legal person which performs the activity to gain profit, an entrepreneur is a physical person who performs the activity to generate a profit and the activity of a free profession, while an individual agriculturist is not an entrepreneur. The Law does not acknowledge the notion of small- and medium-sized enterprises, and, because of that, the same provisions apply to them as to the other enterprises.

The new Law on Business Companies (2008) is a modern regulation, greatly harmonizing with the directives of the European Union company law and, as such, should contribute to the creation of a legal framework complementary the internal market of the EU.

The Law on Business Companies of the Republic of Srpska (Rajčević 2009) is based on the best solutions of modern national law of the surrounding countries, and also of some other countries in Europe and the states in the USA (Illinois), the Statute of the European Company from 2001, OECD Principles of Corporate Governance from 1998 etc. The decrease of bureaucratic obstacles by eliminating the actions of administrative procedures for the purpose of meeting spatial requirements, in which the business activity is performed, in terms of technical equipment, and meeting the requirements of conditions of protection and work and environmental protection, are some of the innovations in the Law on Business

Companies. The simplification of procedures for establishing business companies is harmonised with the Framework Law on Registration of Business Entities in Bosnia and Herzegovina and the Law on Registration of Business Entities in the Republic of Srpska, being also harmonised with the EU community law, and should shorten the procedure for establishing a society to a period consistent with the surrounding countries. The legislator rejected, as anachronous, the term enterprise and accepted a modern, more precise and narrower term, accepted in most surrounding countries, which is "business company", including one of the following types: partnership, limited partnership, limited liability company and joint-stock company.

Entrepreneurship, in the sense of The Law on Development of SMEs of the RS (2013), is an innovative process of creation and development of business ventures or activities and of creation of business success in the market. Entrepreneurial infrastructure presents spatial-technical forms to support entrepreneurship development, with a special emphasis on establishing and developing SMEs.

In recent years, there is a greater emphasis on the commercialisation of university research, especially through the creation of spin-off enterprises (Garmendia and Castellanos 2012). They emphasize the inhomogeneity of the concept of university spin-off enterprises and point out their heterogeneous properties. The suggestions of Garmendia and Castellanos (2012) for the various classifications of university spin-off enterprises are: independent spin-off enterprises, connected spin-off enterprises, with joint investment and as organizational units of universities. Three key approaches are used for differentiating the types of university spin-off enterprises: researchers as entrepreneurs of spin-off enterprises, by the nature of knowledge transfer and the participation of external partners in a new company. These various criteria comprise the phenomena included in the concept of university spin-off enterprises.

The participation of researchers as entrepreneurs (Garmendia and Castellanos 2012) is an important approach, because the profile of an academic spin-off company exploits the opportunity of being connected with universities. Inclusion (Garmendia and Castellanos 2012) of external partners in a new venture is a sign of the quality of such enterprises, and that implies that a spin-off company has successfully passed the evaluation of professional investors and indicates that the transfer of knowledge in a university is at an exceptional level.

The study (Changa et al. 2016) is based on the entrepreneurial university. The concept of Research Ambidexterity (RA) is developed. They came to the conclusion that the development of RA and entrepreneurial universities should be divided into several levels of university, departments and individuals.

The research results (Landry et al. 2010) indicate the existence of three types of knowledge transfer. The first portfolio is made up of complementary activities that are mutually dependent. Another portfolio includes teaching activities and publications that can replace teaching activities. The third portfolio consists of educational activities and other activities independent of teaching activities, i.e. patents, spin-offs, consulting and informal knowledge transfer.

2 Business Zones as Entrepreneurial Infrastructure

Formation and development of business zones is a long-term, planned activity directed to stimulate economic development and employment in the territory of a local community, with the use of adequately equipped facilities and other instruments of support, which enable a more efficient and faster economic and spatial development of enterprises that operate in a zone.

Business zones (The Law on Development of SMEs of the RS 2013) are a form of entrepreneurial infrastructure that offer a structurally arranged and communally equipped space, intended for a harmonised and planned use by a large number of enterprises and entrepreneurs, where the planned and harmonised approach enables a joint use of the space, as well as of communal, administrative, financial, technical and other services—thus realising lower costs of business.

The terms entrepreneurial infrastructure and business infrastructure often have multiple meanings because the development terminology mostly has not been fixed due to its complexity, by legislation and the fact that those are relatively new development mechanisms.

The notion of business zone can define the widest notion of zones in general, which presents a certain area of an infrastructurally equipped building lots that are regulated by spatial-planning documentation, intended for business, i.e. the creation of added value.

Pursuant to the first classification (Medaković 2012), the zones can be classified into four groups:

1. Specialized zones—incubators, technology centres, technology parks, centres for transfer of technologies and zones specialized for certain activities;
2. Industrial zones— areas with a great concentration of industry, predominated by big enterprises;
3. Entrepreneurial—craft zones—areas with a great concentration of small enterprises and entrepreneurs;
4. Agricultural zones—zones founded on soil which is not intended for buildings but is used for agricultural production.

Pursuant to another classification (Medaković 2012), business zones can be classified into the following four categories:

1. Industrial zones are larger zones mostly oriented towards bigger industrial enterprises from similar agricultural sectors, but also the small and medium-sized enterprises (SMEs) related to the mechanism of subcontracting with bigger enterprises. A special category of industrial zones is so-called industrial parks, whose speciality is to have a company as the operator which manages the zone on behalf of one or more owners. In developed countries, the operator can be a public property (public communal enterprises and/or municipality and sate), a public-private partnership or privately owned concern. The aim of creating this form is a more efficient management of the zone and better planned development.

2. Entrepreneurial zones are smaller zones primarily intended for SMEs and entrepreneurs, which have a more favourable support treatment with the aim of faster development, i.e. to invest more in equipment, human resources and working assets, and less in the business premises.
3. Business centres are business zones where business, trading and logistical centres oriented to service activities are most often grouped. The building of business centres is most frequently a private or public-private initiative.
4. Technology parks are the zones directed to high technologies and usually emerge near universities (with technical faculties and institutes). Located in technology parks are usually micro-, small- and medium-sized enterprises based on high technologies, the application of new knowledge and introduction of new practices into the economy.

Besides the named classifications (Medaković et al. 2016a, b), on the basis of strategic importance, industrial zones could be divided into zones of strategic interest that are defined pursuant to various criteria, as projects of special interests, and emerge by an initiative of the government towards the realization bearers, which can be of various levels and legal statuses (from top to bottom) and local zones of municipal or regional importance and oriented towards smaller industrial capacities. Their size is 10–60 ha, depending on the needs and possibilities of the organizer. The initiatives for the development of such zones originate from one or more municipalities (initiatives from bottom to top).

Business zones present special organised business units in which, at one location, well connected with communications, the types of production and service activity based on the principles of cluster organization are developed, with the use of developed infrastructure and accompanying services that reflect their specific industrial characteristics.

Basically, business zones secure the competitiveness for businesses in two ways (Cvijić 2007; Medaković 2012):

The first is related to the possibility of using the effects of integration of similar and related businesses within a zone, resulting in the making of competitive advantages for downstream activities within the zone by means of: access to various sources of raw materials, components, packaging and services; lower transaction costs, because the locations of providers and producers are identical; efficient coordination based on availability and constant exchange of information among the buyers and suppliers in the zone; improvement of innovation processes on the basis of good knowledge of the consumers' needs and of joint work in solving problems; specialisation and efficiency increasing and the application of new technologies in a strategic partnership of enterprises in the zone; firmer integration of providers and related industries in the chains of value of enterprises in the zone, especially in cases when downstream activities are oriented towards international markers; and partnership with related industries in cases when they can service a few enterprises, for example, when distributive enterprises in the zone can distribute the products of a few producers.

The second aspect is related to efficiency offered by the location of the zone for its members by means of: decrease of investment costs for production and business objects; decrease of operational costs of functioning, transportation, maintenance and safety of an object, and the services organized in the zone; and joint use of certain objects (laboratory, copy-room, energy sources etc.).

Business zones appear under various names. Other terms also in use are: industrial park, economic zone, business zone, industrial possession, business zone, artisanal zone, eco-industrial park and some others, but basically they denote what has been subsumed under the notion of a business zone. Regardless of what we name them (Medaković 2012), all of them have two characteristics in common: common location of enterprises oriented to mutual business cooperation and a common managerial structure. They vary one from the other in the sense of type and size. They are most often divided due to the type of investment, i.e. the preparedness for investment, to the green field and brown fields. In the first case, green field, we talk about the creation of business zones at completely new locations, while in the second case, a brown field zone has been created from already used ground and objects in industrial centres. From the viewpoint of Bosnia and Herzegovina, especially significant is the use, or reactivation, of infrastructural capacities of former state enterprises, whose value rapidly evaporates by lack of use. In cases when the reconstruction of existing capacities is more expensive then the construction of new ones, the priority is the first option.

Business zones have certain specificities in relation to free zones, which are, in many countries, one of the instruments for conducting trading policies. Moreover, free zones are specially denoted and arranged areas in one state, where business activities take place under special conditions, mostly with certain benefits related to elimination of duties and taxes for activities directed towards export markets. So, free zones are basically the means of export promotion and of promotion of direct foreign investments in some countries. Besides the stated differences, business zones and free zones also have some similarities, in the sense that both forms of organization are directed towards the building of competitiveness by means of creating a competitive advantages via a certain location of production. Moreover, the free zones in industrially developed countries have lost significantly in their primary meaning of economic oases based on the advantages of a duty-free area and the avoidance of taxation, and they gain importance in creating competitive advantages from the fundamentals of a quality infrastructure of a free zone, high technologies applied in the zone, advantages of specialization, innovations and low transactional costs and other advantages created by the business zones.

The basic differences and similarities of business and free zones (Medaković 2012) are shown in Fig. 1. The similarities are shown in the intersection of squares and differences in each of the squares which present particular forms of zones.

Business zones should be considered as one of the instruments for the realization of new industrial policies that promote many important economic goals. Among the goals, the following ones stand out: restructuring of production; growth of employment; increasing productivity and efficiency in the economy; improvement of the technological level of production and business in general; improvement of

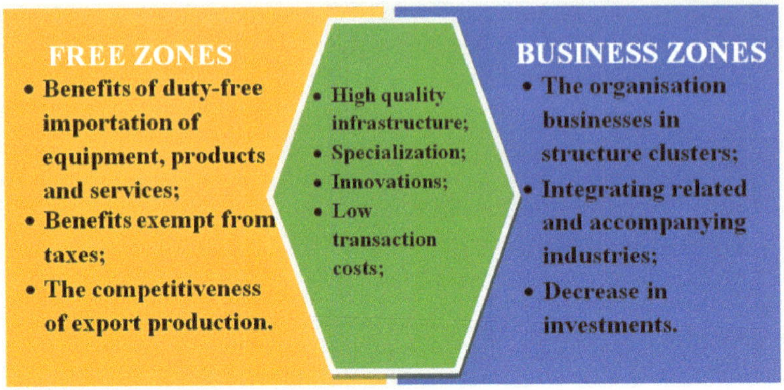

Fig. 1 Similarities of and differences between free and business zones

export and export competitiveness; and development of small and medium-sized enterprises (SMEs).

Establishing competitiveness in a small country, especially in conditions of responsibility for regional and local development, is related to the growing role of small and medium-sized enterprises. Specificity of development based on the promotion of SMEs is related primarily to the need of creating a business environment in which the enterprises will enjoy the conditions for building sustainable competitive advantages. In many elements, the business environment of SMEs exhibits specificities. The enterprises do not have the strength to act independently in big markets because of their fragmentation, so it is logical that, in the business environment, they start to build partnerships and develop cooperation in horizontal and vertical dimensions to promote the cluster-type cooperation.

Namely, what cannot be secured as a desirable business environment in the economy in general, is often achieved within business zones, so that they become attractive for the location of production which means entering into an enhanced phase of competitiveness or means greater efficiency within the arena of competitiveness which marks an economy and guided by factors of development.

Some of the goals (Medaković 2012) for founding business zones can be the following ones: securing long-term conditions for the development of small and medium-sized entrepreneurship and production craftsmanship; long-term decrease and alleviation of the trend of unemployment and support for entrepreneurs to create new job positions, especially within the production activities; stimulation of growth and development of entrepreneurs, especially in terms of development of new products; application of new technologies and support for exports; facilitating communication and support of cooperation among the entrepreneurs within a zone, especially the support for association in realization of concrete entrepreneurial and development projects; and creation of conditions for a transfer of a part of production activities from a cramped town centre to use that space for more suitable and profitable goods.

Economic development (Medaković 2012) of an area is greatly determined by available potentials, i.e. resources, on one hand, and certain factors, i.e. the measures which create a favourable ambient and support to development, on the other hand. Available resources for economic development of a region—contained in: infrastructural capacities (roads, railway); natural potentials; power sources; installed economic capacities; personnel; and geographic location—present a relatively solid basis for the future planned development of the region.

3 Characteristics of Business Incubators

In the practice of small business development, the incubator model deserves special attention. It seems to be a practical approach in the countries of traditional support for the development of small business, and it seems that its perspectives also exist in the transition countries. Business incubators present, as stated by Bošnjak (2011), a contemporary tool for entrepreneurship development in Europe and the USA, and they appear as a response to the recession era and the failure of industrial systems. Most business incubators (Bošnjak 2011) use the existing abandoned infrastructure and brown-field investments for their development. Actually, wherever the surpluses in capacity (empty halls, warehouses, agricultural objects etc.) occur, due to privatisation or other reasons, it is possible to use them to develop incubator types of small business. It is, basically, a flexible method for new business development and support for economic development on a confined, local space. Incubators enable many small enterprises to start their work under the same roof with a favourable lease of a functionally prepared space, joint use of infrastructure, services, and specialized types of equipment. Also, they offer equal opportunities for using certain financial, technical and marketing programs. Incubators, located under one roof, enable multiple combinations of business cooperation among the owners of small enterprises, and, on the basis of that, an efficient use of their available resources in a narrow space.

A business incubator (The Law on Development of SMEs 2013) is a form of entrepreneurial infrastructure whose basic activity is to offer services by putting at the disposal, with or without charge, a business premise, consulting, administrative, technical and other services to newly founded businesses, not older than 5 years.

Among the participants in the creation of incubators (Dostić 2001) may be: social communities; sponsors (associates); donors (contributors); firms as members or subtenants of incubators; and a service organization to perform common tasks for incubator participants. The social community (state, municipality) assists in creating the basic infrastructural conditions, especially regarding the provision of a location and respective capacity for incubators, but also in the creation of a local economic basis for the development of entrepreneurship under the roof of incubators. They also create some wider opportunities for gathering modest private capital to their partnership association for the purpose of establishing enterprises.

State institutions (Dostić 2001), for example, a municipality or any other institution, if they possess a space with potential for an incubator system, may donate it to the members of the incubator, under the condition that the businesses of importance for local economy and households are developed under its roof. Such a space can also be conceded under conditions of a favourable lease, so that the incubator subtenants could initially have favourable conditions to nurture and develop their work. Members of the incubator, or "tenants", as they are sometimes called in America, participate with immediate environment and internal contents of the incubator in order to exploit mutual contacts and businesses with less risk and more business security. The aim of this approach is to provide the numerous benefits offered by the incubator on the basis of a consolidated organization of work, programs and modern solutions, joint equipment, quick project feasibility, the advantage of joint maintenance of buildings etc.

Starting from this, a very significant question arises: what types of enterprises can or should be potential members or subtenants of incubators? That, above all, depends on a determination of what an incubator should contain, for example, only production activity, crafts-service activity, trading activity or, maybe, the competitive freedom among some of them. What should be especially be taken into account is of how and where to find the locations and buildings for such ventures (Dostić 2001). With that aim, abandoned smaller factories or particular production facilities should be discussed, as well as abandoned or unused schools, warehouses, storehouses, business premises, communal houses, buildings which were once used by the military etc. The locations of such buildings can be acceptable if they are near established business centres, in the suburbs of big cities, in small cities, bigger industrial and village settlements, tourist locations etc. As appropriate locations and buildings which can be used and revitalised, the ones that are damaged or destroyed to a certain degree can be used as well, if not too large are the financial resources necessary for their revitalisation.

The rules that have to be established to enter into the incubator of potential business start-ups can be prescribed by law or defined by the main sponsors and initiators of its establishment. Uncontrolled entry by potential incubator occupants into this system would lead to errors, either of a technical or of a marketing nature, and to quick business failure of such enterprises. Also, the potential members of an incubator should get familiar with the detailed concept of its functioning, especially regarding the centralized services, administration and use of common space. Those who do not accept such a concept cannot be the members of the incubator. Entrepreneurs, as candidates for the incubator, should be introduced to all possible mistakes, especially those that would likely be fatal for business results.

In the end, there is no unified concept (Medaković 2012) for, model of or means to develop incubators. There are broad possibilities to establish the contents of work, to design innovative solutions for or opportunities related to urban, suburban or rural position of incubators and the like, and, due to that, the opportunities for specific and untypical types of the business complex. In smaller places, incubators can even be the centres of integrated development of business and employment, and, because of that, every particular case can have individual specifics.

Entrepreneurial incubators gather, in one place, most of the services (Cvijić 2007) that are necessary to an entrepreneur for a faster, successful development of an enterprise. The services are specially related to: offering the services of using business premises, consulting services, the development of human abilities, connections of chamber and institutions and financial services.

The development of incubators as the instruments of economic policy (Cvijić 2007) is especially supported by developed countries, while this cannot be said for countries in transition. The contribution of incubators is reflected by the fact that, when their work well, they significantly decrease the number of enterprise collapses, and they enables enterprises to create new job positions and diversification of production, so that they significantly contribute to the development of the small-business milieu and to local and regional development.

The conveniences and advantages of enterprises that operate in incubators are: use of knowledge and experience of expert and management team of the incubator, as well as the access to knowledge by means of linking into networks on a wider territory; mutual connection of entrepreneurs and opportunities for experience exchange; possibility of additional training and enabling through various programs organized by the incubator; possibility of easier access to financial and investment types of assistance; technical services of incubators; and positive climate and working environment that contribute to inventiveness and innovation of entrepreneurs (Cvijić 2007).

Incubator managers and management (Dostić 2001) are among the most important factors for the future success of an incubator because success depends on their ability to manage to connect on the basis of shared interests with the parties seeking work and the success of the incubator (local government, business associations, enterprises), and, through that, contribute to the position and status of the incubator, realizing a quick development of their members, i.e. the enterprises. For an incubator to be long-term sustainable, it is necessary that the management possesses the ability to create and maintain a positive business environment and business culture, as well as the ability to provide assistance in the fields of marketing and business planning.

Business incubators usually consist of a great number of small business units (usually 10–50 enterprises) (Medaković 2012). There are usually enterprises with these traits: located in one place, mostly within one building; physically separated by room dividers (it is desirable that there is a possibility of moving the dividers if the enterprises have need to take more space); the building in which the enterprises are located should have, at least, common rooms for the incubator manager, business-consulting services, joint meetings, a refreshment room where the participants can spend time together informally. Also, the facility must be equipped with adequate infrastructure: electricity, water, loading and unloading platforms, a sufficient number of telephone lines, parking space etc. The spaces are offered to entrepreneurs under flexible conditions, with low rents, and simple and favourable lease agreements. The incubator is equipped with integral services for offering business-support services to entrepreneurs on the spot, at low prices or for free.

The mentioned services are specially related to training programs through which entrepreneurs can improve their knowledge and skills in specific business areas. There are also the services of business consultants, in the form of advice, business connecting and also the connecting of entrepreneurs with organizations that can provide the capital for the start-up and development of a business.

For the essential nature of the operation and significance of an incubator, the most appropriate definition (Cvijić 2007) states that an incubator is an enterprise that operates for various purposes, in various ways, and it helps through various different forms the founding of new enterprises and their development into profitable independent enterprises.

One of the approaches for classification of incubators incorporates the criteria and aims of their founders (i.e. owners), with possible cases (Cvijić 2007):

- incubator is founded as an enterprise that leases the business premises and makes profit, which is the criterion of the ownership approach,
- incubator is founded as an enterprise which, besides inexpensive business premises, supplies professional advice and helps the development of entrepreneurship, thus manifesting the social goal of supporting the development of small and medium-sized enterprises. This type of incubator must also be financially supported, usually during the first 5 years of its existence.

The actual practice (Cvijić 2007) shows a larger number of entrepreneurial incubators, such as:

- Traditional or classical business incubators that offer their clients an arranged workspace and expert business advice on the spot.
- Administered organized premises are "naked" business incubators that offer to small enterprises, their tenants, only organized workspaces at affordable prices, not giving business advice. It is assumed that the enterprises have already survived the initial period and that they have already developed. This type of incubator enables renting, but pursuant to market prices, both of bigger premises and of consulting and technical services. Such an approach characterizes the US conception of an entrepreneurial incubator.
- Incubators without walls are organizations that do not offer business premises, but only the organized business-support services.
- Incubators of the new economy are highly specialized in the fields of quick-growing sectors, such as high technology and the Internet. The advantage to the founders is the expected benefit from the quick growth of the value of enterprises in which the owners have their share.

Between the traditional business incubators and the business incubators of the new economy, a great number of different organizations can be placed, dealing more or less with providing support for the development of innovations and entrepreneurship, such as: business and innovation centres (BIC), technology centres, knowledge centres, technology-transfer centres, technology parks, science-technology parks etc.

A traditional or classical business incubator provides the business premises under favourable conditions (with accompanying office, secretary and similar technical services) and also provides business services (training and business advice), which comprises the framework of the concept of so-called traditional or classical business incubators. This type of incubator, usually, usually functions to enhance regional and local economic development (through the development of entrepreneurship and creation of new enterprises) and in the creation of new jobs. The most important characteristic of the concept and success of this type of incubator is the fact that they offer quality services and business advice individually to every small enterprise. Without this, such business premises would be administered organized workspaces—only similar to incubators. Their emphasis on socioeconomic functions conditions their character, being non-profit almost as a rule, and it means that national and local authorities and the public sector must have a significant role in their founding and operations. Classical business incubators sometimes more emphasise social components, because it is not rare that they are used as the supporting instrument for the development of entrepreneurship of certain social groups, such as women, returnee immigrants, refugees or national minorities.

The Business Innovation Centre (BIC) is the dominant model of business incubator in the European Union. BICs differ from the US incubator model (Medaković 2012) primarily because of the fact that they give greater significance to consulting and developing entrepreneurs' professional skills, being less targeted at providing premises at lower rents. They are, actually, an improved model of the US business incubator. Thanks to BICs in certain areas, from different sources, by means of various projects and advice to entrepreneurs, local growth and development of services necessary for a success of small and medium-sized enterprises is realised and supported. The concept of business incubators in Europe started to be built in the 1980 s on the basis of the US model, with the aim of alleviating the consequences of breakdown of the large business systems and overcoming the problem of unemployment. It can be said that it demonstrated its justifiability. In the territory of Europe (Cvijić 2007), the most famous institution for help and promotion of BICs operates successfully—namely, the EBN (European Business and Innovation Centre Network) which gathers more than 180 BICs. In the realization of its mission, the EBN supports local sources and the development of abilities, securing the unified activity of target groups in the territories similar to regions so that BICs, in their operations, work pursuant to certain standards of the EU. The incubator personnel, both permanently employed and temporarily engaged (Talent Pool), basically provide their clients with twelve main services shown in the following overview (Cvijić 2007): selection of entrepreneurial ideas; selection of entrepreneurial projects; evaluation of entrepreneurial ideas; evaluation of entrepreneurial projects; training and preparation of entrepreneurs; technical services; marketing consulting; preparation of business plans; rental of space; acquisition of space; rental of services; and entrepreneurial advice.

Technological centres offer a highly suitable form of entrepreneurial infrastructure, in which technologically demanding programs are realized on the basis of consulting and a mediatory role. The centres have incubator-like characteristics because they enable the numerous services and cooperative relationships for enterprises. In a technology centre, there are opportunities for: association of entrepreneurs on the basis of interest in various projects and groups; business of a larger number of enterprises in one place with business infrastructure provided; connections to other consulting institutions and networks; access to the bases of patents; and assistance to entrepreneurs on the formalities for obtaining credit or risk capital. Technological centres usually have their own laboratories and special measurement, testing and control equipment. When they reach a certain mass and level of their own equipment, they can easily transform into a technology park.

A knowledge centre is an organization established with the aim of enhancing research in a certain area. The knowledge centres are usually established by public research organizations, faculty institutes and business organizations. They are usually organized as organizational units.

A technology-transfer centre is a research-development unit organised as a business association.

A technology park unifies, in its structure, the work of technology centres and affirmed medium-sized and big enterprises. There is an emphasis on technology transfer and the cooperation of scientific and research institutions with enterprises. Technology parks enable the renting of larger areas at market prices for prototype production, and it does not impose time limits on enterprises regarding the use of the space. A technology park (The Law on Development of SMEs 2013) is a form of entrepreneurial infrastructure that, within a defined space and with adequate equipment, performs the connection of scientific and research institutions with business subjects for the purpose of technology transfer, application of innovations and development of the economic area in which it is placed. In the countries in transition (Cvijić 2007), due to underdevelopment and the small number of incubators, technology parks work partially as incubators, because the state supervises the space and services, so that the length of stay in a park are limited and only a few enterprises can be accommodated. Technology parks have a significant role in the development of the region because, pursuant to the EU criteria, they are placed in an area where about 300,000 people live, having one or two universities and a few supranational associations (Cvijić 2007). Technology parks are, through joint investments, founded by a country, regions, cities, major enterprises, faculties, banks etc., investing non-refundable assets in the form of land, buildings and money. The members of the management board of a technology park are the most prominent businessmen, representatives of local institutions and representatives of founders and are the basis of further connecting interested parties and securing assistance for the incubator. A technology- park manager is a key person for success, since he/she must be an expert in many business fields and capable of leading a team.

A science park is an institution in which scientifically-oriented people, e.g. researchers, professors, financial experts, consulters and businessmen, are concentrated. At the same time, it is the place that connects science and economy. The basis of the activity of science parks is the application of achievements and innovations in the field of engineering and technology on a commercial basis, and support for the founding and development of enterprises based on knowledge. The most common location of a science park is close to a university, and the founders can vary from a country, municipal and local institutions to banks and chambers of commerce. As stated by Nelles and Vorley (2010), universities are the motors of an economy based on knowledge and also the places where knowledge is produced and exploited. They further address the improvement of the concept of entrepreneurial infrastructure as the analytical framework for understanding the organization dynamics of a modern university and insurance of entrepreneurial evolution within higher education. Some research (Powersa and McDougall 2005) gives useful insights into the planning and performance of technology-transfer activities. Research shows that the degree of industrial R&D, the quality of faculty and venture capital are significant indicators of the technology-transfer effect. They put the accent on universities as motors of economic development and on increasing their engagement in technology transfer in the field of entrepreneurship. Some research (Degroof and Roberts 2004) has paid special attention to the characteristics of academic "spin-off policies", where technology transfer and entrepreneurial infrastructure were weak outside high-tech clusters, indicating a significant influence of academic institutions on the potential growth of spin-off policies.

4 Clusters as a Model of Entrepreneurial Infrastructure

A cluster is a network organization or a group of enterprises coordinated by market mechanisms (Medaković et al. 2016a, b) rather than by chains of commands. A cluster consists of enterprises that are mutually connected, vertically (buyer–supplier) or horizontally (same buyers, technologies, distribution etc.). Competitive branches are grouped into clusters most often.

A cluster (the Law on Development of SMEs) is a form of entrepreneurial infrastructure consisting, in a geographical area, of mutually related business entities that do their business in the same, similar and various activities and, related to them, specializes suppliers, service providers, educational and scientific-research institutions, agencies and other.

In the practice of small-enterprise development, the cluster model deserves special attention. It has proven to be practical, especially in countries that have a tradition of supporting the development of small enterprises. The immediate predecessor to the development of clusters was incubators and, actually, clusters developed quickly in the areas that have had experiences with incubators. The connecting of entrepreneurs, their cooperation, sharing of services and equipment, or their networking into the most common clusters is also, as stated by (Bošnjak 2011), the influence of the work of most incubators. The connecting of business incubators with the academic and research community is also of great significance.

Clusters are presented by Porter (1998) as the basis of the new competitive economy, emphasizing their importance in increasing the competitiveness of enterprises on local and international markets. He defined them as geographic concentrations of mutually related enterprises, specialised suppliers, service providers, enterprises with similar activities and relevant institutions related to them (universities, agencies for standardisation and vocational associations), which mutually compete in some areas but also earn money. A cluster, as a concentration with critical mass (Porter 1998) and extraordinary competitive success in a certain activity, represents a significant characteristic of every country, regional and local economy, especially in economically developed countries.

In countries in transition, clusters have their real future perspectives. It must be kept in mind that the experiences gained in the development of clusters in the European Union can be transferred to local practice without modifications, which would be unavoidable in an uncontrolled development.

The system of clusters (Jojić and Božić 1998) is a global model for the development of small enterprises. They prosper where the development of small business has already reached a significant level and where systematic measures of the country contribute to it. They assist enterprises to develop quickly, to apply modern methods of work and gain the maximum from their market environment under modern management; at the same time, the regional economy achieves competitive advantages in relation to others. Thus clusters connect technologies, industries and, generally, the economies of two regions or even a few regions from a few countries. It can occur within branches (metal, wood, textile, food etc.) in which SMEs develop by using communications, technologies and foreign investments from other regions.

The aim of clusters is to achieve a competitive advantage. One of the relevant factors promoting competitiveness is also the geographic concentration. Finally, besides the geographic (local) factors, legislation, organization and development of financial market have great significance for the development of clusters. The use of knowledge (Medaković 2012) about sound strategies for the development of clusters in the world and on the overall results from their realization contributes so that, in all new cases (either specific or not), good solutions for their creation and development are established.

However, besides the similarities, there are also differences in the development of clusters. They mostly occur due to different degrees of development of regions, levels of governmental inclusion in this model of small enterprises development, strengths of the industrial basis, critical masses of people with entrepreneurial spirit and the general interest of environment factor carriers in general. In developed countries, local and regional government initiate the development of clusters and have had success in that, because of good knowledge of economic processes and of exactly where they are in control, consistent with finding productive measures of support for the realization of the goals (Medaković 2012). In underdeveloped countries and in small countries, ventures of this type are taken at the national level, especially when it is known that local and regional authorities are not ready to support the development of clusters.

5 Review of Existing Forms of Entrepreneurial Infrastructure in the Republic of Srpska

At the beginning of the 1990s, characterized by the breakdown of the former state, the outbreak of civil war, especially in the territory of B&H (The Republic of Srpska), events stopped not only the development of entrepreneurship but also fundamental economic activities. The support for development of small and medium-sized enterprises in the Republic of Srpska gained in importance in 2002, with the adoption of the Program of Small Business Development, and, after that, the Law on Stimulating the Development of Small and Medium-sized Companies was adopted. The adopting of the law created the basis for legislative, institutional and financial help in this area. On the basis of the law, during 2004, there were formed two key agencies: the Department for SMEs and Production Craftsmanship at the Ministry of Economy, Energy and Development of the Republic of Srpska and the Republic Agency for the Development of Small and Medium-sized Enterprises. At the same time, on for local level, local agencies for the development of SMEs were being established. The support the development of SMEs at the local level is also now given by municipal development departments which, together with the above mentioned institutions, maintain support network for the development of SMEs.

Besides the state institutions, the institutes for small business at universities and colleges have a significant role in the support of the SME sector in the USA (Medaković 2012). In the European Union, state institutions also support the work of SMEs, creating a favourable setting for their business. The development of SMEs in the EU occurs at various levels, although the measures are not harmonised and the same for all members, which implies that a successful practice cannot be copied because of the existence of differences in cultures, economic structures and economic potentials.

Every local community or a set of local communities that are connected geographically engages in various activities to improve the conditions of work for SMEs so as to attract a larger number of enterprises to their territory or their region (Medaković 2012). Local communities plays a very significant role, while the task of the country, or the government, is to activate the internal resources, as additional developmental sparks.

A prudent activity of a local community aiming to develop its own infrastructure and entrepreneurial potential and to attract investments can be to adopt a concept for regional development. Local communities must develop attractive environments for capital and enterprises. The establishment of business zones accelerates and simplifies the placement of spatial resources in the function of economic development, investments, growth and employment. Everywhere in the world, business zones present a significant instrument for the stimulation and development of entrepreneurship and general economic growth of an area. They are established on the basis of a clearly expressed interest among the businessman and bodies of local and regional government, with the support of higher levels of government and research-educational organizations, universities and institutes.

Regarding the business incubators, their location is also important, and it is recommended that an incubator serve as an entrepreneurial environment, so it is desirable that they be networked with the university or science sector. Incubators accept (Medaković 2012) primarily firms based on new technologies (hi-tech firms) or production firms based on tested technologies. The emphasis should be put on human factor as the generator of success. Business incubators, as well as business zones, can have a significant, positive impact on the development of the national and local or regional economy.

The entrepreneurial infrastructure (Law on Development of SMEs of the RS (2013)) can be organized in the form of business zones, technology parks, entrepreneurial incubators and clusters. Exceptionally, they can be organized and other forms of infrastructure for the development, promotion and research in this area in accordance with the regulations and requirements of the economy.

Table 1 (IRB RS 2017) provides an overview of the entrepreneurial infrastructure in the Republic of Srpska. This overview contains the forms of organization of entrepreneurial infrastructure clusters and entrepreneurs-business centers: free zones, industrial and business zones, science and technology Parks, and business incubators, as well as their activities and the number of existing forms of entrepreneurship infrastructure.

As regards cooperation with scientific institutions and universities in the Republic of Serbian, there exists several business incubators and one cluster; they are (IRB RS 2017):

1. Business Incubators—Innovation Centre Banja Luka operates as a foundation, which has the following stakeholder's structure: Ministry of Science and Technology of the Republic of Srpska, Atene Prosjekledlse on behalf of Ministry of Foreign Affairs of Government of Norway, City of Banja Luka, University of Banja Luka, University of East Sarajevo and RARS—Republic Agency for the Development of Small and Medium Enterprises. The Innovation Centre Banja Luka contains enterprises: Business Incubator, Training and Conferences.
2. Business incubator NGO "Krajina" Banja Luka, located on an area of ten hectares of land planted with perennial medicinal herbs (seedlings garden formed in partnership with the city). The business incubators are agriculture and forestry activities, and owned by City of Banjaluka. Project leaders: NGO Krajina"—Business Incubator, City of Banjaluka, City of East Sarajevo, RS municipalities, RS Government, RS Institute of Agriculture and Banjaluka and Centre for Rural Development and Improvements. A business incubator has been implementing a programme of organized support for the development of agriculture—production of organic food, fattening of breeding cattle and initiation of small farms (interviews conducted with 1734 households in the city territory, education, advisory groups and establishment of agro-exchange—electronic supply and demand bids for agricultural products).

Table 1 Review of existing forms of entrepreneurship infrastructure in the Republic of Srpska (IRB RS 2017)

Entrepreneurial infrastructure		Activities	Numerical strength
Cluster	Existing clusters	Agriculture and forestry	5
		Processing and products of wood and cork, cellulose paper and paper products	6
		Energy, gas and water production and supply	1
		Food and beverage industry	1
		Other	2
	Planned	Food and beverage industry	1
		Agriculture and forestry, food and beverage industry	1
	Cooperatives	Food and beverage industry	4
		Agriculture and forestry, food and beverage industry	6
Entrepreneurs-business centers	Free zones		3
	Industrial and business zones		96
	Science and technology parks		1
	Business incubators	Agriculture and forestry	1
		Construction, furniture production, production of metal products, excluding machinery, production of rubber and plastic products	1
		Other, processing and products of wood and cork, cellulose paper and paper products, production of leather and leather products	1
		Other	6

3. Business Incubators university business incubator—in creation phase. It will be cited in the municipality East Ilidža.
4. Existing clusters, Solar Energy Cluster, whose is activities Energy, gas and water production and supply. Leader cluster is Faculty of Mechanical Engineering Banja Luka with members “Topling Factory” Prnjavor, “Energomont-Bemind” Banja Luka and “Koming pro” Gradiska.

6 Conclusion

As the result of determination and preparations for membership in the EU, in Bosnia and Herzegovina (The Republic of Srpska), there is currently a process of reforms with the aim to harmonise the existing legal system with the legal system of the EU. The current determinations are to harmonise the legal framework, which is being built with the EU and the Act on Small Business, providing the conditions for the SMEs sector to be at the top of economic development priorities. These regulation will result in healthy legal frameworks for the founding and status of companies, including the fields of taxation, the availability of financial resources and the orderly cessation of the work of a company. The principles valid in the process of reforms are: decentralised organisation and modest administration and making of conditions for providing the possibility of realization of business interests and initiatives of interested parties, i.e. the participants in the SME sector.

It is very important to make efforts in understanding particular notions of different forms of entrepreneurial infrastructure. There are no precise and unified definitions, so the matching is not compulsory among the notions in the various countries.

General characteristics of entrepreneurial infrastructure (business zones, incubators and clusters) are presented, as well as their role and importance in terms of providing conditions for the creation and development of small and medium enterprises.

A review of the literature comes to the conclusion that particular notions are named differently, but, in principle, are the same shape entrepreneurial infrastructure.

The paper presents the existing state of the entrepreneurship infrastructure in the Republic of Srpska, with the numerical strength of individual forms, as well as their activities.

From the point of view of activities and the impact of scientific institutions and universities in the Republic of Srpska and their participation in establishing and supporting the work of the incubator, the results are not satisfactory. We have two incubators that exist, which includes the universities and scientific institutions, as well as a cluster in which participates as a founder the Faculty of Mechanical Engineering Banja Luka. We also have a university business incubator, which is in the process of formation.

On the other hand, traditional, passive teaching on entrepreneurial matters is largely present at the universities. In the future, scientific institutions and universities should be more involved in the establishment of incubators that should enable, through advice, training and consultancy, the business success of future entrepreneurs, as well as their creation of companies that would have a chance to survive in the market.

References

Bošnjak S (2011) Business incubators as generators of entrepreneurship development. Practical Management. J Theory Appl Manag 2(2):51–56

Changa Y-C, Yangb PY, Martinc BR, Chid H-R, Tsai-Line T-F (2016) Entrepreneurial universities and research ambidexterity: a multilevel analysis. Technovation 54:7–21. doi:10.1016/j.technovation.2016.02.006

Cvijić M (2007) Research of conditions and opportunities for the development of entrepreneurship in the transition process. Doctoral thesis. FTS Novi Sad 2007

Degroof JJ, Roberts BE (2004) Overcoming weak entrepreneurial infrastructures for academic spin-off ventures. MIT IPC working paper IPC-04-005

Dostić M (2001) Managing small businesses. Sarajevo 2001

Garmendia BJM, Castellanos RA (2012) Types of spin-offs in a university context: a classification proposal. Cuadernos de Gestión 12(1):39–57

Heger D et al (2011) The effect of broadband infrastructure on entrepreneurial activities: the case of Germany. Centre for European Economic Research (ZEW). Discus si on paper no. 11-081. http://ftp.zew.de/pub/zew-docs/dp/dp11081.pdf

IRB RS (2017) URL: www.irbrs.net. I approached the review of data, day: 01.02.2017

Jojić R, Božić R (1998) Entrepreneurship for the new century. Srpsko Sarajevo, Priština

Landry R, Saïhi M, Amara N, Ouimet M (2010) Evidence on how academics manage their portfolio of knowledge transfer activities. Res Policy 39(10):1387–1403. doi:10.1016/j.respol.2010.08.003

Medaković V (2012) Research of conditions for development of small and micro enterprises and development of entrepreneurial infrastructure model. Doctoral thesis. Faculty of Mechanical Engineering. East Sarajevo

Medaković V, Marić B, Gojković R (2016) Cluster as a model organization for small and medium-sized enterprises. 3rd International scientific conference" Conference on mechanical engineering technologies and applications" COMETa 2016. East Sarajevo, Jahorina 2016, pp 517–522

Medaković V, Marić B, Moljević S, Miljanović M, Gojković R (2016) Entrepreneurial infrastructure—characteristics of business zones. In: VI International conference industrial engineering and environmental protection 2016 (IIZS 2016). Zrenjanin 2016, pp 8–13

Nelles J, Vorley T (2010) Constructing an entrepreneurial architecture: an emergent framework for studying the contemporary university beyond the entrepreneurial turn. Innov High Educ 35:161–176. doi:10.1007/s10755-009-9130-3

Porter ME (1998) Clusters and the new economics of competition. Harvard Business Review, New York, pp 77–78

Powersa BJ, McDougall PP (2005) University start-up formation and technology licensing with firms that go public: a resource-based view of academic entrepreneurship. J Bus Ventur 20:291–311

Rajčević M (2009) The law on business companies of the Republic of Srpska. Materials with counseling. Banja Luka 2009

The Law on Enterprises of the Republic of Srpska (2006) Gazette RS. Number 24/98. 62/02. 38/03. 97/04 and 34/06

The Law on Business Companies of the Republic of Srpska (2008) 01-1882/08. Gazette RS. Number 127/08

The Law on Development of SMEs of the Republic of Srpska (2013) 01-916/13

Characteristics of Support Programmes for Setting Up Spin-Offs in Spanish Universities

José María Beraza-Garmendia and Arturo Rodríguez-Castellanos

Abstract This paper seeks to validate the models of support programmes for setting up spin-offs found in the literature analysing the differentiating characteristics of the various types of programmes existing in Spanish universities. To do this, a survey was conducted among managers of these programmes to obtain the necessary information. Secondly, the variables characterising these programmes were established by performing a factorial analysis. Thirdly, based on these variables, several different types of programme were identified in Spanish universities using cluster analysis. Finally, the differentiating characteristics of each of the types found were identified by means of Anova and logistic regression analysis. Four types of programme were found with differences in terms of human resources, experience, commitment, proactivity, selectivity, type and number of spin-offs created and the involvement in their management. Two successful models were identified from among the four types of programmes found. This study confirms the importance of R&D activity as a determinant of the results of support programmes for setting up spin-offs in universities. However, the suggestion in the literature that programmes usually begin with a policy of low selectivity was not confirmed.

Keywords Knowledge transfer · Support programmes for setting up spin-offs · Entrepreneurship · University spin-offs

1 Introduction

Transfer of research results to the market from universities in general and, more particularly, through the creation of knowledge-based firms has become increasingly important as a source of development and competitiveness (European Commission 2002; Gómez et al. 2008; Karnani 2013; Shane 2004), bringing with it

J.M. Beraza-Garmendia (✉) · A. Rodríguez-Castellanos
Department of Financial Economics, University of the Basque Country, UPV/EHU, Plaza de Oñati 1, 20018 San Sebastián, Spain
e-mail: josemaria.beraza@ehu.es

J. Peña Dopazo and F. Zivic (eds.), *Supporting University Ventures in Nanotechnology, Biomaterials and Magnetic Sensing Applications*,
DOI 10.1007/978-3-319-61237-9_3

a proliferation of spin-off support programmes in universities (Berbegal-Mirabent et al. 2012; European Commission 2004a; Festel 2013; Helm and Mauroner 2007; Lundqvist 2014; Shane 2004; Swamidass 2013). Nonetheless, there are considerable differences in the way the various programmes operate and are structured, in terms of their aims, strategies, functions and organizational structures (Cáceres and Aceytuno 2015; Cesaroni and Piccaluga 2016; Clarysse et al. 2002, 2005; Degroof 2002; European Commission 2002; Gïlsing et al. 2010; Mustar and Wright 2010; Roberts and Malone 1996; Wright et al. 2007). In addition, the activities and organisational structures of the newly founded programmes have not been clearly identified (Cesaroni and Piccaluga 2016; Clarysse et al. 2005; Heirman and Clarysse 2004). For this reason, various models of academic spin-off support programmes have been proposed (Cáceres and Aceytuno 2015; Cesaroni and Piccaluga 2016; Clarysse et al. 2005; Degroof 2002; Fini et al. 2011; Roberts and Malone 1996; Wright et al. 2007).

These studies focus on universities in the U.S. and some European countries, such as France, Belgium, the United Kingdom, the Netherlands, Germany, Italy and Ireland. It may therefore be of interest to validate the models identified in these studies by analysing the situation in other countries.

The main aim of this work is therefore to validate the models of support programmes for setting up spin-offs found in the literature, analysing the differentiating characteristics of the various types of programmes to be found in Spanish universities and identifying those that seem to be more successful.

Using data taken from a survey of people in charge of university programmes for supporting the creation of spin-offs, a univariate and multivariate statistical analysis was performed. Firstly, the technique of factorial analysis was applied to identify the most significant variables explaining the characteristics of these programmes. Secondly, various different types of programmes in Spanish universities were identified by cluster analysis. Using a one-factor analysis of variance (Anova) and a logistic-regression analysis, the differentiating characteristics of each of the clusters found were then described.

The paper comprises an introduction and six other sections. The second section reviews the literature on types of university spin-off support programmes, identifying the various different models suggested by different authors. The third section explains the reference model for the study, the questionnaire used, the sample selected and the way the data were collected. In the fourth section, as previously explained, a univariate and multivariate statistical analysis is performed. The fifth section provides a descriptive statistical analysis to characterise each of the models identified. The sixth section offers a logistic-regression analysis to highlight the main differentiating features. Finally, the seventh section summarises the principal conclusions of the study.

2 Literature Review

This article is particularly concerned with the support programmes set up by universities to promote the creation of academic spin-offs, namely, with the measures adopted by the universities to support the creation and development of this type of company. Based on the classification established by Pirnay (2001), it adopts the perspective of university management, the so-called "university level" of analysis, to use the classification established by Pirnay (2001).

Two main groups of work can be distinguished in the literature in this field: some that analyse the organizational and institutional factors determining the creation of spin-offs; and others that examine the process of creating the spin-offs.

Among the first group, studies that have looked at organisational aspects highlight the role of the technology-transfer units found in some universities as a mechanism of support for setting up spin-offs (Hague and Oakley 2000; McDonald et al. 2004) and identifying the success factors of these units (Algieri et al. 2013; Berbegal-Mirabent et al. 2012; Cesaroni and Piccaluga 2016; European Commission 2004b; Lockett et al. 2003; Lockett and Wright 2005; Rodeiro et al. 2008, 2010; Van Looy et al. 2011; Vinig and van Rijsbergen 2009).

Within this first group, some studies have focused on the determining role of institutional behavior, in this case of the university, on the spin-off activity. According to these authors, universities with a culture of supporting activities to market their research results will obtain better results from spin-off creation. They include Calvo et al. (2017), Franklin et al. (2001), Gómez et al. (2008), Guerrero and Urbano (2012), Hayter and Rooksby (2016), Muscio et al. (2016), Polt et al. (2001), Rasmussen and Wright (2015), Siegel et al. (2003), Thursby and Kemp (2002) and Uyarra (2010).

The second group of studies analyses the process of creating the spin-offs. Several authors have used a linear approach to describe the spin-off process and its various stages (Clarysse et al. 2005; Fernández-Alles et al. 2015; Pirnay 2001; Roberts and Malone 1996; Shane 2004; Vohora et al. 2004); there is, however, no consensus as to the number of stages, what they should be called and at what point the process begins.

Finally, within this second group of studies, some authors examine the characteristics of the various different university policies of support for spin-off ventures. For example, Roberts and Malone (1996) find that the two main dimensions of a policy of economic valorisation through the creation of spin-offs are *selectivity* and *support* and that universities have only two viable support policies: *low selectivity/low support* and *high selectivity/high support*.

According to these authors, in favourable environments, the university can adopt a passive strategy, benefitting from an existing entrepreneurial culture in which the creation of spin-offs is common and venture capital abundant; in such cases, it will therefore adopt a low-selectivity/low-support policy. In contrast, in unfavourable environments where spin-offs are rare and venture capital is scarce, the university has to play a more active role: fostering entrepreneurial culture, searching for marketable research results, getting involved in the selection process and supporting the spin-off creation and development process. It will therefore adopt a policy of high selectivity/high support.

Degroof (2002) analyses how an environment that is unfavourable to entrepreneurship affects the type of spin-offs created in academic institutions. In this type of environment, he distinguishes two models for spinning off, depending on the type of institution: the specialist research institutions and the universities. The author concludes that the process adopted by specialist research institutes involves pursuing a high-selectivity/high-support policy, whereas the process identified in universities involves a low-selectivity/low-support policy, thus contradicting Roberts and Malone's conclusions. However, the author notes this process is not static, but becomes increasingly more sophisticated as institutions learn from their experience. He therefore deduces that, in an unfavourable environment, it is difficult to establish a high-selectivity/high-support policy from the outset. Implementing such a policy requires considerable resources and skills, which may not exist in the universities; moreover, implementing them involves, among other factors, bringing about a considerable cultural and structural change. In unfavourable environments, therefore, a university can start out in a position of low selectivity/low support and gradually move towards a position of greater selectivity and support. However, it is likely that not all universities will be able to make this transition, and even if they can, it remains to be seen at what speed they can do so.

For their part, Clarysse et al. (2002, 2005) and Wright et al. (2007) distinguish four categories of support programmes for setting up spin-offs in universities: *low selective* or *self-selective*, *supportive*, *resource-deficient* and *skills-deficient*.

In the *low-selective* model, there is an attempt to stimulate entrepreneurial initiative, with less attention paid to economic or financial potential. This means that the universities' main activity consists of raising awareness and searching for opportunities. The aim of this model is to create as many spin-offs as possible, not only technology-based spin-offs but others based on skills developed in the university.

In the *supportive* model, management of intellectual property and preparation of the business plan are essential. Financing, public and private, is also needed to enable the projects to develop in the initial stages. In this model, the spin-offs are an alternative option for obtaining value from the development of technology. Therefore, only technological spin-offs are usually supported. The aim is not to create as many spin-offs as possible, but to create spin-offs with an ambition for growth.

The *resource-deficient* group is characterised by a scarcity of resources of all kinds, experience and know-how of staff, contact networks to develop all the activities required to support the creation of spin-offs and a lack of support from the research institution's management team. Because of these deficiencies, such programmes are classed as being of weak support.

The *skills-deficient* group includes support programmes for setting up spin-offs that have the necessary resources but lack the abilities to integrate the resources needed to generate the required skills.

In summary, the literature review concludes as follows: firstly, there is more than one model of support programme for setting up spin-offs; secondly, the two main dimensions of a policy of economic valorisation through the creation of spin-offs are selectivity and support; thirdly, environment-specific conditions determine the objectives to be achieved, the resources to be used and the activities to be carried out by the universities if they are to succeed; and fourthly, many universities do not have the resources to establish a high-support and high-selectivity policy; even if they can, they may require a certain period of time to develop the necessary skills to utilise these resources efficiently.

The model used as a starting point for the empirical study, questions asked in the questionnaire on which this study is based and the process employed for gathering and processing the information are presented below.

3 Methodology

3.1 Starting Model

Several authors have analysed the process and the various stages of creating the spin-offs using a linear approach (Clarysse et al. 2005; Degroof 2002; Fernández-Alles et al. 2015; Pirnay 2001; Roberts and Malone 1996; Shane 2004; Vohora et al. 2004). There is, however, no consensus as to the number of stages, what they should be called and at what point the process begins.

Design of the empirical study is based on a model of the linear process[1] of valorisation via spin-off proposed by Pirnay (2001),[2] which considers that the university's involvement is essential in the early stages but becomes less relevant as the process advances. This does not mean that the university cannot be involved in the final stages, but that there are other agents (financial institutions, science and technology parks, business incubators etc.) that can favor their development more effectively and efficiently (Rasmussen and Wright 2015).

3.2 Questions

This empirical study used information taken from a survey on a series of variables related to resources and results, activities, organisational structure and the university's relations with the spin-offs.

---

[1]Although the division of the spin-off process into six successive stages aids understanding of the process, it should not be taken to imply our complete acceptance of the linearity of the innovation process, as the real situation is more complex. The various phases in the process do not occur in an ordered consecutive sequence, one after another, but normally involve relations of reciprocity and interdependence.

[2]Pirnay (2001) distinguishes six successive stages: promoting an entrepreneurial culture, searching for and detecting ideas, evaluating ideas, developing projects, setting up companies and consolidating value creation.

The questions included in the questionnaire refer to the model of the linear process of valorisation via spin-offs. This model sets out the various stages a university needs to consider in supporting the creation of spin-offs, with particular emphasis on stages and activities at which the university's direct intervention can be decisively important. These can be grouped analytically into four basic stages: promotion of entrepreneurial culture, search for and detection of ideas, evaluation and valorization of ideas and the creation of spin-offs. To these four stages, we have added an additional section on general information.

The boundary between these four stages is sometimes blurred, since there is a certain overlap between the different areas. Nonetheless, we feel that this division is useful as an analytical outline for deciding on the questions to be asked and indicating what aspects of the support programme they reflect.

3.2.1 General Information

Support programmes for setting up spin-offs are usually developed by a specific organisational structure (Cáceres and Aceytuno 2015; Cesaroni and Piccaluga 2016; European Commission 2004b; Lockett and Wright 2005). For this reason, as well as the identifying data, three additional questions were also included in the survey: the type of body performing the spin-off set-up support activities; the number of people in that body; and the activities related to this type of company carried out at the university.

3.2.2 Promotion of Entrepreneurial Culture

The encouragement of an entrepreneurial culture amongst university personnel is a preliminary step for creating spin-offs in unfavourable environments for entrepreneurship (Cáceres and Aceytuno 2015; Henry et al. 2005; Klofsten 2000; Pirnay 2001). For this reason, four questions were included in the survey: the university's level of commitment to promoting entrepreneurial culture; the actions being taken to promote this culture; the number of people that will benefit from these activities; and the degree of success obtained.

3.2.3 Search for and Detection of Ideas

Commercially exploitable ideas deriving out of university research do not normally emerge spontaneously (McDonald et al. 2004; Rasmussen and Wright 2015; Shane 2004; Siegel et al. 2003). The questionnaire included four questions related to the search for and detection of ideas: the proactivity of the university; the type of actions carried out to search for and detect ideas; the source of the entrepreneurial ideas and the degree of success obtained.

3.2.4 Evaluation and Valorisation of Ideas

It is necessary to determine whether the ideas detected meet a series of conditions for viable commercial exploitability (Wright et al. 2004). Likewise, the university and the people who have engendered the ideas must support them finally in becoming a spin-off (Rasmussen and Wright 2015; Vohora et al. 2004). Nine questions were included in the survey on this section: the use of a specific methodology for evaluating and exploiting ideas; the use of external personnel in technological assessment; the use of external personnel in market assessment of the idea; the profile of these personnel; who assumes the leadership in promoting the spin-off; the role generally taken in the spin-off by the research group from which the idea originated; the number of exploitable ideas detected in a 1-year period; the percentage that are favourably evaluated; and the percentage that ultimately lead to the creation of a spin-off.

3.2.5 Creation of Spin-Offs

The spin-offs created can be of several types and be more or less successful (Bessière et al. 2015; Cáceres and Aceytuno 2015; Clarysse et al. 2005), and the university can have various types of links with them (Lockett et al. 2003). Nine questions were included in the survey for this reason: number of spin-offs created in the last 5 years; type of spin-offs created; average length of time required for the process of creating the spin-off; types of financial resource used; university's stake in the spin-off's capital; university's involvement in its management; survival rate of this type of company; percentage of firms that fail before 3 years and, finally, the year the spin-off creation support programme was implemented at the university.

3.3 Sample and Data Collection

The study population comprises European universities in general, and Spanish universities in particular, that perform some type of activity involving spin-off creation.[3]

To identify the study population and select the sample, a list of European universities was first identified in Google, classified by countries; secondly, their respective websites were consulted; thirdly, the body responsible for spin-off set-up-support activities was ascertained; and, finally, the persons in charge of these bodies were identified (name, position, telephone number and e-mail address).

[3]This study is part of a broader study evaluating the degree of implementation of university spin-off support programmes in Europe, the measures applied, the main problems and the success achieved.

Altogether, a total of 329 universities were identified in Europe (not including Spanish universities). These universities were sent a letter of presentation by e-mail, inviting them to fill out the online questionnaire. Replies were received from 67 universities, representing a response rate of 20%. In the case of Spain, virtually all universities with a Technology Transfer Office (TTO), the great majority of which were public universities, received the letter of presentation. Thirty-five replies were received, representing a response rate of 58%. Thus, 389 letters of presentation were sent, replies were received from 102 universities, so the response rate was 26%.

Any replies with a given number of incomplete answers or those considered atypical were then eliminated. Finally, a database was obtained including a total of 65 universities, of which 23 were in Spain and 42 elsewhere in Europe. The rate of complete replies was therefore 17% for the sample as a whole, 38% for Spanish universities and 13% for other European universities.

The data thus obtained were statistically analysed using univariate and multivariate techniques. Initially, the technique of factorial analysis was applied to identify the most significant variables explaining the characteristics of these programmes. The different types of programmes were then identified in Spanish universities by cluster analysis. Next, using a one-factor analysis of variance (Anova) and a logistic-regression analysis, the differentiating characteristics of each of the clusters found were described. Finally, the principal findings were summarised.

4 Multivariate Analysis

4.1 Representative Substitute Variables

To determine the most significant variables explaining the characteristics of the support programmes for setting up spin-offs, a factorial analysis was made with the database drawn from the information obtained from all universities.

Using this technique, the factors were first identified. To do this, the *Principal Components Analysis* data-reduction method was used. Likewise, the *latent root criterion* was used to determine the number of factors to be extracted. The contributions of each variable to the different factors were then calculated, and the variables that contributed most to each one were selected, in order to identify the variables that most accurately described the university spin-off support programmes and use them in the subsequent cluster analysis.

Forty-seven variables were originally used in this study, all the quantitative variables in the survey; by applying the *latent root criterion*, the final solution chosen was that formed by 16 factors. In order to improve the solution, a Varimax

Table 1 Most significant variables, classed by stages in the spin-off exploitation process

General information	• Number of people in the body
	• Relative importance of promotion of entrepreneurial culture
	• Relative importance of search for and detection of ideas
Promotion of entrepreneurial culture	• Success in the promotion of entrepreneurial culture
Search for and detection of ideas	• Importance of monitoring projects undertaken by research groups
	• Importance of monitoring business design/project competitions
	• Relative importance of "Others" in originating ideas
Evaluation and valorisation of ideas	• Relative frequency of postgraduate-Ph.D. students as leaders in promoting spin-offs
	• Relative frequency of external personnel hired as leaders in promoting spin-offs
	• Appropriateness of research groups providing technological consultancy to spin-offs
	• Percentage of ideas detected over a 1-year period that are positively evaluated
	• Percentage of positively evaluated ideas that lead to the creation of a spin-off
Creation of spin-offs	• Percentage of knowledge-based spin-offs
	• Percentage of "Other" spin-offs
	• University's involvement in management of the spin-off
	• Percentage of spin-offs that expire in under 3 years

rotation was used. This solution preserves 78.652% of total variability. The variables with the greatest load for each factor were then selected, as being representative of each of the factors. In this way, it was possible to group into 16 variables the original quantitative variables that are intended to represent different aspects corresponding to the various stages of support for setting up spin-offs. Table 1 shows these variables.

4.2 *Typology of Support Programmes for Setting Up Spin-Offs in Spanish Universities*

Once the representative substitute variables had been obtained, the Spanish universities were classified using the cluster-analysis technique, from the 16 most significant variables identified in the previous sub-section. The distance used was the *Euclidean distance squared*, following standardisation of the variables by transforming them into *Z scores* with mean 0 and standard deviation 1. Hierarchical clustering and *Ward's method* were used to form the clusters.

There is no standard statistical method for determining the final number of clusters, and the clustering coefficient criterion does not give clear results in this case, since it does not experience significant change when the number of clusters varies. We therefore opted for obtaining between two and four different cluster arrangements and checking whether there were significant differences between the clusters obtained, by using the one-factor analysis of variance (Anova), taking the variable containing the information on the cluster to which each university belongs as an independent variable or factor and each of the variables included in the analysis (the 16 most significant variables identified in the previous sub-section) as dependent variables and checking whether there were significant differences between the clusters obtained.

The results obtained in the Anova analysis show that there are a number of variables with significant differences that are practically the same for the two-, three- and four-cluster arrangements. The four-cluster arrangement was therefore chosen, since it allows greater differentiation between the programmes, without proving excessive. In the four-cluster arrangement, significant differences were found in the following six variables: "number of people who are part of the body", "importance of monitoring business design/project competitions", "percentage of ideas detected over a one-year period that are positively evaluated", "percentage of positively evaluated ideas that lead to the creation of a spin-off", "percentage of knowledge-based spin-offs" and "university's involvement in management of the spin-off".

5 Characteristics of Support Models for Setting Up Spin-Offs in the University

Having identified the types of support programmes for setting up spin-offs in Spanish universities and the representative substitute variables with significant differences, we will try to identify in more detail the characteristics of these clusters, as representatives of different models or types of spin-off creation support, comparing and contrasting them to highlight their similarities and differences.

To this end, all the quantitative variables with significant differences from the four types of support programmes included in the survey will be used, as well as two new variables that are not included: return on R&D and number of patents. The reason for including these two variables is as follows: given the somewhat subjective nature of cluster analysis and, in order to ensure the validity and practical relevance of the solution obtained, it is appropriate to include variables that are not used to form the cluster but whose value is known to vary from one another. This is the case of the two variables chosen. Moreover, both are relevant in this respect, as we show below.

Amongst inputs related to the university business creation process, R&D activity is one of the factors most used and most often cited in the literature. The variable included in the analysis was, specifically, the return on R&D for academic year 2002–2003 in each of the universities, expressed as a percentage of return on R&D of the university with the highest income.

Patent applications by universities are also an indicator of university R&D results and their commercial orientation. A relationship might therefore be expected between the number of patents and the business creation process in the universities, although not all academic spin-offs are based on patented knowledge. However, in some cases, the number of patent applications filed by the universities varies greatly from 1 year to another.

For this reason, the variable used in the analysis was the average number of patent applications filed in the Spanish Patent and Trademark Office during the period 2000–2005 in the Spanish universities.

Table 2 shows the means of all the quantitative variables with significant differences obtained in the survey for the clusters of spin-off support programmes.

Taking these variables as a reference point, the profiles of each of the four types of programme identified are as follows:

Type 1 is made up of eleven universities characterised by: they have with a relatively small level of R&D activity that have spin-off programmes started up recently and a small number of people devoted to spin-off support; their commitment to promoting an entrepreneurial culture is limited, and therefore a small number of people are involved in their activities; in searching for and detecting ideas these universities are not very proactive; they give little importance to monitoring business design/project competitions and the number of ideas they detect is therefore limited; they give a positive rating to and spin off a small percentage of the ideas detected; they generate few spin-offs, most of which are technology-based; and they have little involvement in their running. Figure 1 shows the characteristics of Type 1.

Type 2 has strong similarities with the previous type and, to a lesser extent, with Type 4. It includes six universities with a relatively low degree of R&D activity that started up spin-off programmes some years ago and have a small number of people devoted to this work; their commitment to promoting an entrepreneurial culture is likewise limited; however, they are quite proactive in searching for and detecting ideas; in addition, they place quite a high degree of importance on monitoring business design/project competitions; however, the number of ideas detected is very limited; unlike the previous type, they give a positive rating to and spin off a large percentage of the ideas detected; they generate an intermediate number of spin-offs, with a significant proportion of knowledge-based ones, but they do not intervene in their running. Figure 2 shows the profile of Type 2.

Type 3 is formed by two universities with a relatively high degree of R&D activity that have spin-off programmes started up quite a few years ago, and have a

Table 2 Means of all variables with significant differences in clusters

Variables	1	2	3	4	Total
Number of people who are part of the body[a]	1.09	(−) 1.00	(+) 4.00	(−) 1.00	1.30
University commitment to the promotion of entrepreneurial culture	3.18	(−) 2.83	(+) 5.00	4.00	3.39
Number of people that benefit from these activities	(−) 1.55	1.67	(+) 4.50	3.25	2.13
Proactivity in searching for and detecting ideas	(−) 3.00	3.33	(−) 3.00	(+) 4.00	3.26
Importance of monitoring business design/project competitions[a]	(−) 2.45	3.50	(+) 5.00	4.00	3.22
Number of ideas detected over a 1-year period	1.55	(−) 1.00	1.50	(+) 3.00	1.65
Percentage of ideas detected over a 1-year period that are positively evaluated[a]	1.64	(+) 3.17	(−) 1.50	2.50	2.17
Percentage of positively evaluated ideas that lead to the creation of a spin-off[a]	(−) 2.00	(+) 3.67	(−) 2.00	2.25	2.48
Percentage of technology-based spin-offs	(+) 83.45	62.17	(−) 60.00	68.00	73.17
Percentage of knowledge-based spin-offs[a]	(−) 14.18	37.67	(+) 40.00	32.00	25.65
University's involvement in management of the spin-off[a]	(−) 1.09	(−) 1.00	(−) 1.00	(+) 2.25	1.26
Year in which the university began activities to support spin-offs	(+) 2002	1999	(−) 1995	1999	2000
Average number of spin-offs created between 2000 and 2004	(−) 1.57	3.33	3.40	(+) 9.73	3.61
Average number of patents	(−) 6.23	6.92	(+) 21.50	14.00	9.09
% R&D	29.64	(−) 25.50	58.00	(+) 66.50	37.43

The cluster marked with a (+) or (−) has the highest or lowest mean for the corresponding variable
[a]Representative substitute variables with significant differences between the clusters

very large number of people dedicated to supporting these programmes and to promoting an entrepreneurial culture and therefore their activities benefit a large number of people; in searching for and detecting ideas,[4] these universities appear

[4]This statement needs to be qualified, since this is the cluster that gives most importance to the activities of monitoring the research groups' projects and competitions of ideas/creation of spin-offs and the one that has been most successful in searching for and detecting ideas.

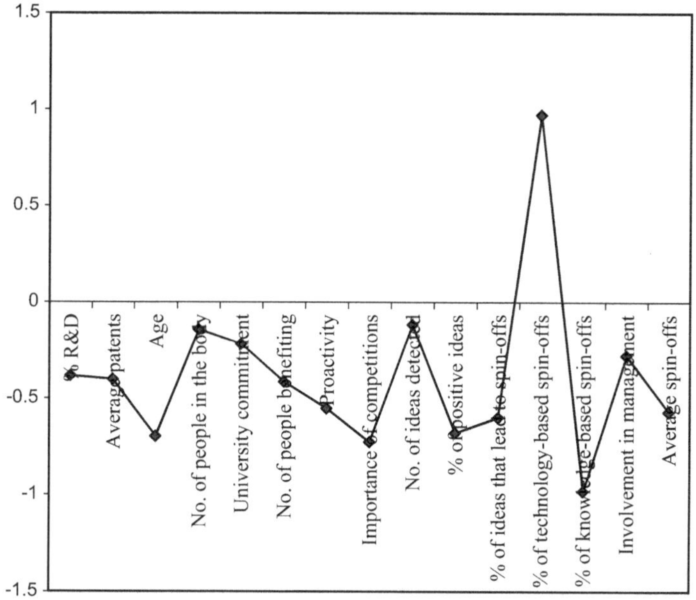

Fig. 1 Profile of Type 1

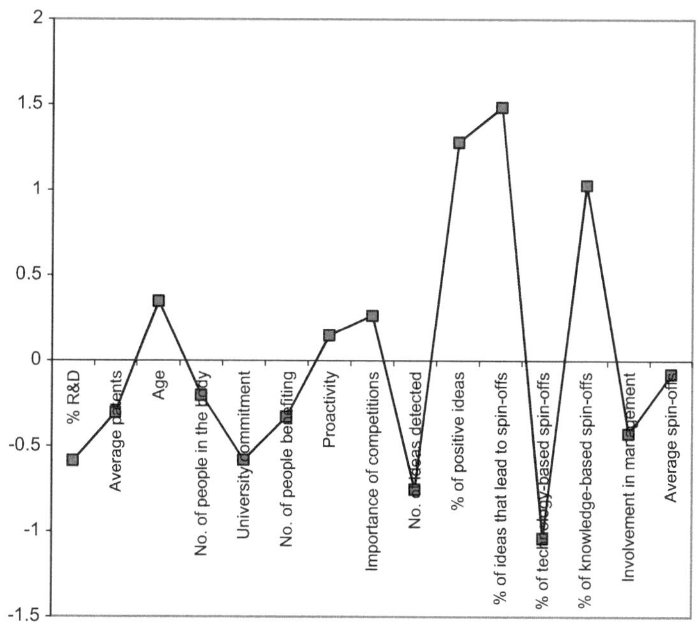

Fig. 2 Profile of Type 2

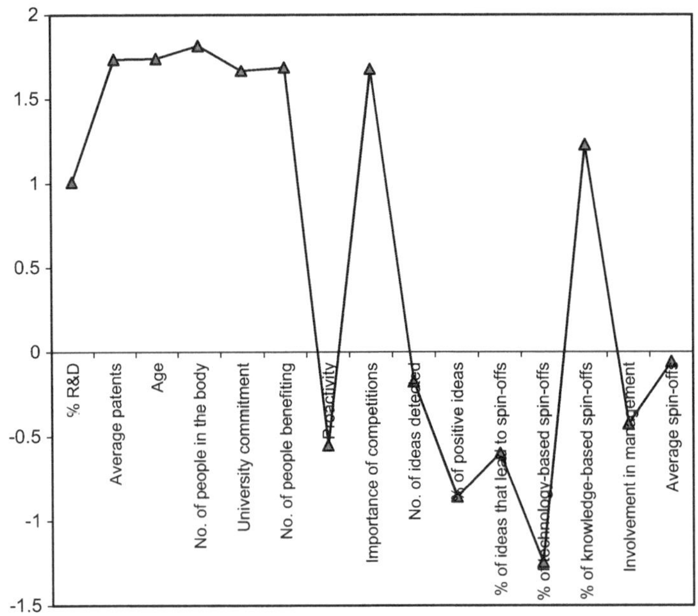

Fig. 3 Profile of Type 3

not to be very proactive; they give great importance to monitoring business design/project competitions, but the number of ideas they detect is relatively small; they give a positive rating to a small percentage of ideas detected and, likewise, spin off a small percentage of the ideas that are positively evaluated; these universities generate an intermediate number of spin-offs, which are mostly knowledge-based; and they do not get involved in their running. Figure 3 shows the profile of Type 3.

Finally, Type 4 is made up of four universities with a relatively high degree of R&D activity that have spin-off programmes started up some years ago, although they have a small number of people devoted to them; their commitment to promoting an entrepreneurial culture is also high, and therefore a large number of people are involved in their activities; these universities are very proactive in searching for and detecting ideas and give great importance to monitoring business design/project competitions, with the result that the number of ideas they detect is relatively high; they give a positive rating to an intermediate percentage of ideas detected and, likewise, spin off an intermediate percentage of the ideas that are positively evaluated; these universities generate a large number of spin-offs, many of which are knowledge-based; and they are involved in their running. Figure 4 shows the characteristics of Type 4.

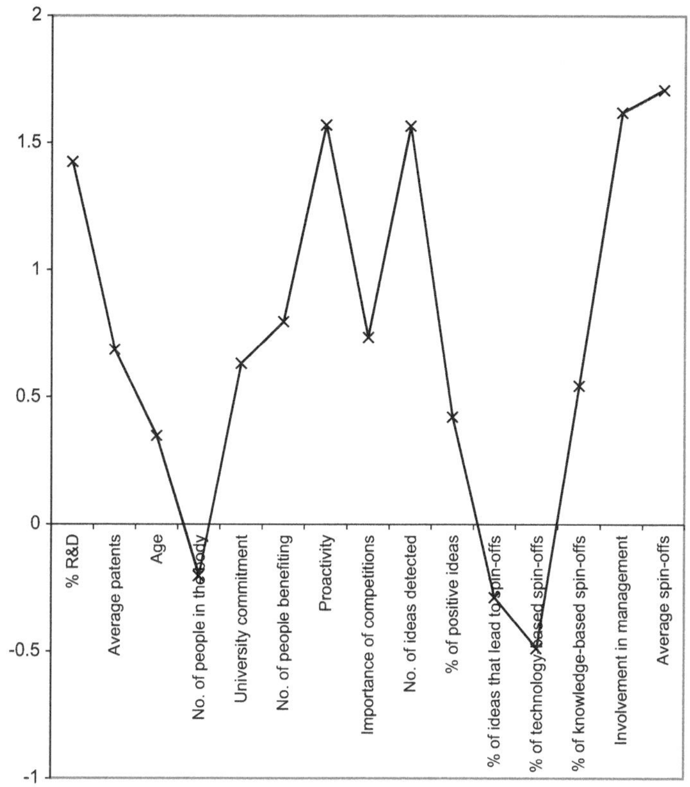

Fig. 4 Profile of Type 4

6 Most Significant Characteristics Differentiating Support Models for Setting Up Spin-Offs in Universities

Having identified the features distinguishing support models for setting up spin-offs in the university, logistic-regression analysis was used to determine whether there were any statistically significant differences between types 3 and 4 and the others and which of these variables best quantified the differences between these two groups of universities. For these purposes, the 15 variables included in the above characterization were used.

Table 3 presents the main statistics of the *logistic-regression analysis*. The value of the likelihood ratio (−2 *log of the likelihood*) falls, which demonstrates that the overall fit of the model improves with the step. Similarly, it is possible to refute the hypothesis that there are differences between the observed probabilities and those forecast by the model, since the *Hosmer-Lemeshow chi-square* statistic is not significant, making it possible to confirm the goodness-of-fit of the final model.

Table 3 Values of the −2 log likelihood ratio, Hosmer–Lemeshow Test and Press's Q statistic for the logistic-regression analysis

Groups of the regression	Step	−2 log of the likelihood	Hosmer–Lemeshow		Press's Q	
			χ^2	Sig.	Value	Sig.
Types 3 and 4	0	26.435	0.014	0.906	12.5652	0.01
Types 1 and 2	1	17.083				

Table 4 The logistic regression analysis: variables and principal results of the model

Variables	Coefficients	Wald	Sense of difference
University's involvement in management of the spin-off	3.219**	5.769	Greater in Types 3 and 4

Significance ** $p < 0.05$

Finally, *Press's Q statistic* makes it possible to conclude, with a significance level of 1%, that the prediction success is better than that obtained randomly.

Once it had been determined that the *logistic regression model* and the precision in the classification were statistically significant, the results were interpreted by examining the relative importance of each independent variable of the model in the differentiation of the groups. For this purpose, the *coefficients of the model* and their significance were analysed. The significance of each coefficient was assessed using the *Wald statistic*. In addition, the sign of the coefficient of each independent variable was studied in order to interpret the *model* obtained. The positive sign of the coefficient indicates that the increase in the corresponding variable raises the probability that a university is allocated to the group with Code 1. The negative sign of the coefficient, on the other hand, indicates that an increase in the corresponding variable raises the probability that a university is allocated to the group with Code 0 or, inversely, that it reduces the probability that a university is assigned to the group with Code 1.

Table 4 shows the variables included in the *model*, their respective coefficients and their levels of significance.

The *logistic-regression model* obtained reveals that only one variable, 'university's involvement in the spin-off's management', can be seen to be positively associated with the dependent variable. Given that this is dichotomous, types 3 and 4 with Code 1 and types 1 and 2 with Code 0, this implies that this variable occurs to a greater extent in types 3 and 4.

Consequently, types 3 and 4 mainly differ from types 1 and 2 in having more university involvement in management of the spin-off.

7 Conclusions

In this paper, we have set out to identify the different models of support programmes that exist in universities in Spain. This enabled us to reach a series of conclusions, which are summarised below.

The results obtained in Spanish universities confirm the existence of different types of support programmes for setting up spin-offs, in terms of experience, resources, university commitment, proactivity, selectivity and number of spin-offs created; variables determining their characteristics; and results according to the literature. Therefore, universities pursue various different support policies, utilise various different creation processes and generate spin-offs in various numbers and with different characteristics, as various authors have indicated (Clarysse et al. 2002, 2005; Wright et al. 2007).

Type 3 follows a policy very similar to the *supportive* model proposed by Clarysse et al. (2002, 2005) and Wright et al. (2007), whose aim is to create companies with economic potential and a growth ambition. Therefore, Type 3 appears to be a successful model.

Likewise, Type 4 follows a policy very similar to the model of *low selectivity* proposed by Clarysse et al. (2002, 2005) and Wright et al. (2007), whose aim is to generate the greatest number of companies possible. For that reason, such universities attempt to stimulate entrepreneurial initiative, paying less attention to their economic or financial potential. This means that the main activity of the universities is raising awareness and searching for opportunities. Type 4 appears to be another successful model.

In contrast, Type 1 shows that it is difficult to establish a policy of high proactivity/high selectivity/high support from the outset; nonetheless, the fact that numerous universities pursue such a policy appears to contradict Degroof's assertion that universities will initially be obliged to start with a low-selectivity/low-support policy.

Finally, Type 2 appears to show that some spin-off support programmes lack resources or competencies, apparently confirming Clarysse's conclusions.

At the same time, the typology of the spin-off support programmes identified in this work enables us to make some recommendations for improving the least successful models.

Universities included in Type 1 appear to be clear on the support policy they wish to pursue, but lack the resources and competences needed to put them into practice. Therefore to improve their results they need to strengthen their competencies and devote a greater quantity of resources to their programmes through training and hiring of specialist personnel, as well as by establishing collaboration networks with external agents specialising in each of the activities in the process. One option for small universities, with limited resources, may be to participate in joint spin-off programmes with other universities.

Meanwhile, Type 2 programmes must prioritize the establishment of the support policy they wish to pursue. Universities that adopt such programmes seem to be very proactive and not very selective, but it is really the lack of resources and competencies that leads them to pursue this policy. For this reason, they must first of all establish the support policy they wish to apply, and it is therefore essential to have prior commitment from the university management.

Finally, the results obtained in this paper enable us to offer two additional observations.

With the exception of one case, Types 3 and 4—i.e., the "successful models"—have the highest return on R&D and number of patents. This result is in consonance with the literature, which sees R&D activity as one of the factors related to the business creation process in universities; at the same time, it also shows that patent applications by universities are an indicator of the results of university R&D and its commercial orientation.

The three technical universities included in the cluster analysis are classified in clusters 3 and 4, which are considered the most successful. This is not surprising given that because of their expertise in the fields of science and engineering, these universities are more involved in the technological development of the country and have greater capacity to transfer technology to society.

References

Algieri B, Aquino A, Succurro M (2013) Technology transfer offices and academic spin-off creation: the case of Italy. J Technol Transf 38(4):382–400

Berbegal-Mirabent J, Sabaté F, Cañabate A (2012) Brokering knowledge from universities to the marketplace. The role of knowledge transfer offices. Manag Decis 50(7):1285–1307

Bessière V, Gomez-Breysse M, Messeghem K (2015) Les determinants de la croissance des essaimages académiques. RIPME 28(3/4):37–63

Cáceres FR, Aceytuno MT (2015) Academic spin-offs incubation strategies: the case of the Andalusian region. Cuadernos de Gestión 15(2):113–142

Calvo N, Rodeiro-Pazos D, Fernández-López S (2017) Science and technology parks (STPs) as accelerators of knowledge-intensive business services (KIBS). A case study. Int J Bus Glob 18 (1):42–57

Cesaroni F, Piccaluga A (2016) The activities of university knowledge transfer offices: towards the third mission in Italy. J Technol Transf 41(4):753–777

Clarysse B, Lockett A, Quince T, Van de Velde E (2002) Spinning off new ventures: a typology of facilitating services. Institute for the Promotion of Innovation by Science and Technology in Flanders, Brussels

Clarysse B, Wright M, Lockett A, Van de Velde E, Vohora A (2005) Spinning out new ventures: a typology of incubation strategies from European research institutions. J Bus Ventur 20 (2):183–216

Degroof JJ (2002) Spinning off new ventures from research institutions outside high tech entrepreneurial areas. Dissertation, MIT

European Commission (2002) University spin-outs in Europe. Office for Official Publications of the European Communities, Luxembourg, Overview and good practice

European Commission (2004a) Improving institutions for the transfer of technology from science to enterprise. Best Project ITTE 1.11/2002. European Commission, DG Enterprise

European Commission (2004b) Technology transfer institutions in Europe. An overview. European Commission, DG Enterprise

Fernández-Alles M, Camelo-Ordaz C, Franco-Leal N (2015) Key resources and actors for the evolution of academic spin-offs. J Technol Transf 40(6):976–1002

Festel G (2013) Academic spin-offs, corporate spin-outs and company internal start-ups as technology transfer approach. J Technol Transf 38(4):454–470

Fini R, Grimaldi R, Santoni S, Sobrero M (2011) Complements or substitutes? The role of universities and local context in supporting the creation of academic spin-offs. Res Policy 40 (8):1113–1127

Franklin S, Wright M, Lockett A (2001) Academic and surrogate entrepreneurs in university spin-out companies. J Technol Transf 26(1/2):127–141

Gïlsing VA, van Burg E, Romme AGL (2010) Policy principles for the creation and success of corporate and academic spin-offs. Technovation 30:12–23

Gómez JM, Galiana DR, Mira I, Verdú AJ, Sancho J (2008) An empirical approach to the organisational determinants of spin-off creation in European universities. Int Entrep Manag J 4:187–198

Guerrero M, Urbano D (2012) The development of an entrepreneurial university. J Technol Transf 37(1):43–74

Hague D, Oakley K (2000) Spin-offs and start-ups in UK universities. Committee of Vice-Chancellors and Principals of the Universities of the United Kingdom, London

Hayter CS, Rooksby JH (2016) A legal perspective on university technology transfer. J Technol Transfer 41(2):270–289

Heirman A, Clarysse B (2004) How and why do research-based start-ups differ at founding? A resource-based configurational perspective. J Technol Transf 29(3/4):247–268

Helm R, Mauroner O (2007) Success of research-based spin-offs. State-of-the-art and guidelines for further research. RMS 1(3):237–270

Henry C, Hill F, Leitch C (2005) Entrepreneurship education and training: can entrepreneurship be taught? Part I. Educ Train 47(2):98–111

Karnani F (2013) The university´s unknown knowledge: tacit knowledge, technology transfer and university spin-offs findings from an empirical study based on the theory of knowledge. J Technol Transf 38(3):235–250

Klofsten M (2000) Training entrepreneurship at universities: a Swedish case. J Eur Ind Train 24 (6):337–344

Lockett A, Wright M, Franklin S (2003) Technology transfer and universities´ spin-out strategies. Small Bus Econ 20(2):185–200

Lockett A, Wright M (2005) Resources, capabilities, risk capital and the creation of university spin-out companies. Res Policy 34(7):1043–1057

Lundqvist MA (2014) The importance of surrogate entrepreneurship for incubated Swedish technology ventures. Technovation 34(2):93–100

McDonald L, Capart G, Bohlander B, Cordonnier M, Jonsson L, Kaiser L, Lack J, Mack J, Matacotta C, Schwing T, Sueur T, van Grevenstein P, van den Bos L, Vonortas NS (2004) Management of intellectual property in publicly-funded research organisations: towards European guidelines. Office for Official Publications of the European Communities, Luxembourg

Muscio A, Quaglione D, Ramaciotti L (2016) The effects of university rules on spinoff creation: the case of academia in Italy. Res Policy 45(7):1386–1396

Mustar P, Wright M (2010) Convergence or path dependency in policies to foster the creation of university spin-off firms? A comparison of France and the United Kingdom. J Technol Transfer 35(1):42–65

Pirnay F (2001) La valorisation économique des résultats de recherche universitaire par création d'activités nouvelles (spin-offs universitaires): Propositions d'un cadre procédural d'essaimage. Dissertation, Université du Droit et de la Santé - Lille 2

Polt W, Rarner C, Gassler H, Schibany A, Schartinger D (2001) Benchmarking industry science relations: the role of framework conditions. Sci Public Policy 28(4):247–258

Rasmussen E, Wright M (2015) How can universities facilitate academic spin-offs? An entrepreneurial competency perspective. J Technol Transf 40(5):782–799

Roberts EB, Malone D (1996) Policies and structures for spinning off new companies from research and development organizations. R&D Manag 26(1):17–48
Rodeiro D, Fernández S, Rodríguez A, Otero L (2008) La creación de empresas en el sistema universitario español. Universidad de Santiago de Compostela
Rodeiro D, LA FernándezS Otero, Rodríguez A (2010) Factores determinantes de la creación de spin-offs universitarias. Revista Europea de Dirección y Economía de la Empresa 19(1):47–68
Shane S (2004) Academic entrepreneurship. University spinoffs and wealth creation. Edward Elgar Publishing Limited, Cheltenham
Siegel D, Waldman D, Link A (2003) Assessing the impact of organizational practices on the productivity of university technology transfer offices: an exploratory study. Res Policy 32(1):27–48
Swamidass PM (2013) University startups as a commercialization alternative: lessons from three contrasting case studies. J Technol Transf 38(6):788–808
Thursby JG, Kemp S (2002) Growth and productive efficiency of university intellectual property licensing. Res Policy 31(1):109–124
Uyarra E (2010) Conceptualizing the regional roles of universities, implications and contradictions. Eur Plann Stud 18(8):1227–1246
Van Looy B, Landoni P, Callaert J, van Pottelsberghe B, Sapsalis E, Debackere K (2011) Entrepreneurial effectiveness of European universities: an empirical assessment of antecedents and trade-offs. Res Policy 40(4):553–564
Vinig T, van Rijsbergen P (2009) Determinants of university technology transfer—comparative study of US, Europe and Australian universities. Available. http://ssrn.com/abstract=1324601. Accessed 30 Jan 2015
Vohora A, Wright M, Lockett A (2004) Critical junctures in the development of university high-tech spinout companies. Res Policy 33(1):147–175
Wright M, Birley S, Mosey S (2004) Entrepreneurship and university technology transfer. J Technol Transf 29(3/4):235–246
Wright M, Clarysse B, Mustar P, Lockett A (2007) Academic entrepreneurship in Europe. Edward Elgar Publishing Limited, Cheltenham

Launching University Ventures—Policies, Financing and Strategy

Nikola Makojević and Vladimir Milovanović

Abstract One of the main tasks of institutional framework creators is cultivating adequate productive enviroment for university ventures development. The university ventures should be specifically fertile for research and development commercialization and support and for positioning universities as profitable units within the educational system. The idea is to create supportive programs where research activities can migrate to the actual business environment, produce desirable products and/or services, compete on the market and generate profit.

Keywords University venture · Entrepreneurship · Venture capital · Profit oriented activities · Research commercialization

1 Setting the Stage—University Environment Creation for Spurring Entrepreneurship

The main challenges that economies face nowadays are reaching dynamic growth rates, opening new job posts and becoming more competitive within global markets. Bearing in mind that there is intensive competition in global markets, decision makers of economic policies must be able to develop instruments that will spur knowledge-based entrepreneurship. To achieve this goal, it is necessary to work on constant improvement of the educational system and forge closer ties with the economy. In this context, each state should create an environment in which universities will be transformed from traditional, research-based institutions to modern ones based on commercial laws and technology transfer. In other words, this means that it is necessary to create so-called entrepreneurial universities. Entrepreneurial universities work towards raising internal efficacy and making closer and productive

N. Makojević (✉)
Faculty of Economics, University of Kragujevac, Kragujevac, Serbia
e-mail: nmakojevic@kg.ac.rs

V. Milovanović
Faculty of Engineering, University of Kragujevac, Kragujevac, Serbia

J. Peña Dopazo and F. Zivic (eds.), *Supporting University Ventures in Nanotechnology, Biomaterials and Magnetic Sensing Applications*,
DOI 10.1007/978-3-319-61237-9_4

ties with industry. As for internal efficacy, it refers to efficient management of educational and research activities, staff selection policy, financial management and creation of curricula that are based on labor market demands. Internal efficacy at a university must be based on the main economic principle stating that limited resources must be employed in order to satisfy the maximum of all existing needs. Making closer and productive ties with the economy, which can be regarded as a type of external efficacy, refers to maintaining constant contact with the economy, including the collection of data on the existing needs and dissemination of the possibilities and capacities that universities possess. Nonetheless, often universities neglect external efficacy, thus creating a deep gap between academia and industry.

Entrepreneurial universities must be the torch bearers when it comes to implementing activities to promote entrepreneurial spirit across the internal activities related to the definition of their mission, goals, and the strategy of their presence in the educational markets. Nowadays, in the circumstances of intensive competitiveness in educational markets, entrepreneurial universities are visible through their promotion campaigns, intensive communication with potential consumers and through their diversified ranges of services, surpassing those related to mere education. Such universities, beside undergraduate, graduate and doctoral studies, offer services for professional development, even to those without a diploma from a formal educational institute, but with practical experience. Apart from education and professional development, such universities offer the services of applied research and analysis, model development and product and service testing, including starting their own companies which act as competitors in the market.

Recently, there has been a number of serious studies (Abreu and Grinevich 2013; Dohse and Walter 2012; Etzkowitz 2004; Colombo et al. 2010) conducted in order to examine the state of the arts in educational systems, as well as to propose measures to be taken in the years to come. The studies have shown that it is necessary for universities develop curricula that will foster interaction, collaboration and co-operation with public and private organizations in order to strengthen, generate and exploit knowledge.

In developing an entrepreneurial university, one must bear in mind that it takes maintaining the parallel processes of education and research and entrepreneurial activities. Entrepreneurial universities can be seen as a pre-requirement for commercialization of research, technology transfer and the creation of enterprises based on applied research (Djokovic and Souitaris 2008). This can exert a positive influence on the dynamical growth rate and development, increasing economic competitiveness, as well as a nation's export potentials in the global market. Entrepreneurial universities accomplish their social goals by creating a work force with developed entrepreneurial spirit, i.e. a work force equipped with modern knowledge and skills enabling its members to start their own entrepreneurial ventures, thus reducing the pressure on labor market (Baldini et al. 2006).

Establishing the pre-requirements for development of entrepreneurial universities was common to a number of studies (Debackere and Veugelers 2005; Eun et al. 2005; Falck et al. 2012; Goethner et al. 2012), whose results can be summarized as follows:

(a) Formal factors (those related to): the way universities are organized and structured; the degree of liberty for making entrepreneurial decisions by respective faculties, institutes or individuals within a university; the programs of support for the development of entrepreneurship; and the curricula and training organized so as to foster entrepreneurship.
(b) Non-formal factors include those whose influence is difficult to measure, but whose influence is evident; the most frequently mentioned non-formal factor refers to the attitude of the academic community towards entrepreneurship or towards the development of an entrepreneurial university.
(c) Resources: these factors include human capital, financial capital and the size of the market.
(d) Factors which determine the current status of a university: these factors include the current position and the status of a university in terms of its co-operation with industry, its experience with projects related to industry, its membership in professional associations and the degree to which the university is recognized by the economic community.

Given the factors stated above, it becomes clear that creating an entrepreneurial university in one country poses a serious challenge for a university, as well as for the state. Namely, when it comes to formal factors, a state must pass laws which will enable respective units within a university to function independently, without compromising the functioning of a university as a whole. Simple as it may sound in theory, this has proven to be quite a challenge in creating an entrepreneurial university. The biggest problem is to motivate researchers to become entrepreneurs without compromising educational processes and research activities. On the other hand, when it comes to non-formal factors, it is up to the university to raise awareness of the need for commercialization of the research conducted within the university, as well as of the need to ensure technology transfer, thus creating closer ties with industry. The biggest obstacle in creating an entrepreneurial university refers to the fact that the majority of the teaching staff holds the view that research is intended for science and raising awareness of the research results, and is therefore not intended for commercialization and generating profits (Goktepe-Hulten and Mahagaonkar 2010).

The third group of factors, those related to resources, may be seen as the first step when analysing the potentials in developing,, entrepreneurial spirit“, because, when there are no such resources, it is pointless to make plans related to development of an entrepreneurial university. The greatest limitation refers to researchers themselves, in the circumstances of their inadequate familiarity with entrepreneurship (Haas and Park 2010). If this is the case, it may suggest that a university will have to extend the time required to develop entrepreneurship at the university level. The university faces the need to select adequate strategies for enhancing "entrepreneurial spirit" in its researchers, some of which can be as follows:

(a) "Entrepreneurial spirit" of researchers can be enhanced in collaboration with researchers experienced in entrepreneurship, which may call for hiring external experts and allocating funds for this process.

(b) "Entrepreneurial spirit" of researchers can be enhanced through defining new conditions to be applied in recruitment of the future researchers, requiring that they are experienced in entrepreneurship or in industry.

Nowadays, universities are facing serious challenges (Henkel 2007) related to their need of becoming more entrepreneurship-oriented with a clear and intensive cooperation with industry when it involves solving practical problems. This task becomes even more urgent when one bears in mind the circumstances of limited financial resources coming from public funds, stressing the need to find additional sources of financing. The main idea is to look beyond the traditional borders of understanding what a university is, i.e. institutions intended solely for education and research, and to move further in exploiting their potential in commercializing research, technology transfer and offering services to industry. To make this possible, it is required to transform universities and motivate researchers so as they could become entrepreneurs and profit-oriented businessmen.

2 The Start-up Phase in Creating a University Venture

The beginning of any business venture can be regarded as a challenge, but this is definitely true for researchers who should endeavour to create enterprises that will commercialize the research conducted in universities or institutes. The initial stage is crucial not only in creating university ventures, but also in creating companies other than those created as a result of the need to commercialize scientific research (Markman et al. 2008). Before they start creating a business venture, researchers should be prepared for the challenge of being in business-related circumstances. Every beginning takes time, effort and resources, and for this reason, involvement in a venture should result in returns and making profits. Since researchers are often inexperienced in creating business ventures, they should prepare well. The preparation stage evolves in several steps: finding experienced entrepreneurs who will provide information on all the aspects of initiating the venture; becoming a member of professional networks and associations; and assessing one's own knowledge of a particular venture. Other peoples' experience is of particular importance to people who are aiming at commercializing their own research, and who learn about other peoples' success or failure. In this way, thanks to primary sources of information, researchers become familiar with the dangers and challenges associated with venture lunching.

Involvement in professional networks gives opportunity for better understanding of market affairs, general economic conditions, legal regulations, available financial sources and the forecasting related to research business-venture products or services. The assessment of their own knowledge can take the form of a self-assessment using a questionnaire including questions related to starting the venture, from the registration stage, through organization, accountancy standards, to financial commitments to the state, i.e. taxes. Once they have filled out a questionnaire, researchers should get in touch with an experienced entrepreneur to help them in assessing the responses. In this way, researchers are able to get an objective negotiate of their knowledge and ability to plan, in a timely and resourceful fashion, their business activities. It is essential that researchers take criticism coming from experienced entrepreneurs in a constructive way, and to act accordingly, in order to avoid situations resulting in poor performance in the process of launching a university venture.

2.1 *Knowing the Industry*

The initial stage in development, from research to a product, requires the analysis of the industry related to the university-venture launching (Lam 2007, 2010). Knowing the industry is essential because the collected data can help a researcher when making a final decision, whether to get involved in the venture or not. Knowing the industry implies analysing competition, analysing the movements (growth or decline) in the industry, the state-of-the-art technology, profits rates, competitors' strength, obstacles etc. All these pieces of information are useful to the process of making a decision related to university-venture launching, i.e. they provide answers as to whether one should get involved in a venture in order to enter into a certain market. Should the answer be positive, it is necessary to further analyse the industry, gathering information from both primary and secondary sources in order to create a strong foundation for making decisions.

One of the most common problems in launching university ventures is the problem of asymmetric information, i.e. the absence of true information on the state of affairs and prospects in a certain industry. Such a situation is inherent in the circumstances of a non-existent relationship between a university and industry. Hence, researchers may be in possession of insufficient or misleading information.

In countries, such as the UK or USA, universities are very much involved in industrial development and raising the competitiveness of the economy as a whole. Europe, with the exception of Germany, is lagging behind due to traditional beliefs that science plays the role in raising the quality of life of the society as a whole, and, at the same time, due to slow responses to the demands coming from the industry in terms of collaborative research. The program Horizon 2020 is an endeavour to enhance the co-operation between universities and industry in EU countries, and thus to create conditions for improving entrepreneurship at universities through stronger market orientation and commercialization of research.

2.2 *Market Analysis*

When it comes to initiate a university venture, researchers must analyse whether it is possible to commercialize research, to become a player in the market and to make profit on sales. This is a key moment, because it can be regarded as a turning point when, if the response is positive, success can be achieved. Markets work on the basis of supply and demand, so if a product meets or creates demand, it can be expected that the product can become a commercial success. This is where researchers experience problems, because they often fail to recognize the market needs when they work on solving a research problem. When they develop a solution, it is often too late to negotiate the potential value of the solution, and the researchers' conviction is often hard to be rationalized, as well as it is rarely economically rational. University-venture launching must be based on rational, economic assumptions and analyses; in order to avoid the potential danger of wasting financial and time resources, it should be based on market-analysis information.

The market analysis assumes collection of data and data analysis with regard to:

(a) The current state of affairs in the industry: who the key market players are; whether the industry is growing, stagnating, or declining; how intense the competition is; what legislation in this field is; the level of technological development.
(b) The prospects of the industry: potential obstacles to the growth of the industry; development of the legislation related to the filed; new players in the market.
(c) Available financial sources: what the main sources of funding (banks, stock exchanges, venture capital, business angels) are.
(d) The current level of cooperation between industry and academia.

Data collected through such analysis can be relevant because they are gathered from both primary and secondary sources. In the process of making decisions related to launching university ventures, one most not disregard data collected through non-formal contacts and the experience of other players in a given industry. Secondary data comes from publically available sources, and they create a reliable foundation for starting a research. Given the fact that they are in possession of not only publically available data, universities can be considered to have an advantage, because they often have sole access to the researchers, so that it can be concluded that they may be much better informed than entrepreneurs themselves.

2.3 *Organizing a University Venture*

Following the analysis of a given industry, launching a university venture moves to the operationalization stage, which evolves in terms of formal organization to take the shape of a spin-off, start-up, or SME. To a large extent, the form of organization is imposed by the state and the university regulations related to commercialization of research (O'Shea et al. 2005). Regulatory solutions influence the motivation of researchers to go for commercialization. Namely, if the rules envisage a modest remuneration to be paid out to researchers, it is unlikely that they will be motivated to take part in a university venture, which adds to their day-to-day activities and workload. If, on the other hand, the institution decides to give over the majority of profits to the researcher in charge, it is likely that the researcher will feel motivated to get involved in the commercialization of research. However, some research results suggest that there is a danger to quality of teaching and research in cases in which institutional resources are disproportionately employed to achieve commercial success.

If it is assumed that a researcher is interested and that satisfactory balance between regular activities and commercialization has been achieved, one must consider the manner in which the university venture is organized and structured, as this will influence the way a research is recognized in the market. There are several possibilities such as forming spin-offs, start-ups, or SMEs based on a partnership between a university and a researcher, but it can also be based on total ownership on the part of the researcher with an obligation that a share of the profits should be directed to the university, or it can be based on forming a new enterprise in cooperation with existing companies willing to finance development and expansion of the product in the market.

Once the initial problems in organizing a university venture have been overcome, the researcher can focus on the process of commercialization. When it comes to commercialization, i.e. market positioning, the university venture should opt for strategies to attract consumers, as well as for strategies for entering the market (Nosella and Grimaldi 2009; Powers 2003).

2.4 Creating Market Position

To achieve a successful market position, university ventures should look for an adequate answer to the following question: who are my consumers? This is often the question researchers find difficult to answer due to their lack of market experience, since the invention or development of a product/service does not guarantee its commercial success. If there is no market for the product/service, it will not survive in the market and will become another example of poor understanding of market functions.

To put it into other words, researchers should analyze potential consumers and the position of their product/service in the current supply-demand chain before they decide to enter the market. A product/service can complement the existing supply, it can also meet demands in a new way, or it can create demand, such as in the case of mobile phones, social networks etc. When they analyze consumers, researchers need to investigate their current status, the way in which their own product/service can satisfy their needs, make estimates of consumers' loyalty to the existing producers, the level of consumers' income and its structure, as well as their willingness to embrace new products/services. Such analysis is necessary for targeting consumers who need to be attracted and form the initial market. Such an initial market serves as a pool for data collection on the product/service; it ensures additional capital for investment and marketing; and it enables researchers to make necessary corrections with respect to their competitors. In the starting stage of the venture's presence in the market, it is important to understand that it takes time to attract customers, especially when it comes to industries where consumers are loyal to certain brands, or when it comes to industries with a large number of competitors.

Once the first transactions take place in the market, the university venture should start analysing results, correcting errors, and repositioning itself in the market. There are several ways to conquer the market, one of them being by aggressive performance that quickly conquers the market by competitive pricing, better service, and better quality of products/services. This is possible in branches which are intensive in technological terms and which do not require large investments in promotion and distribution, but which can be introduced to the market using modern means of communication.

A university venture can opt for a niche strategy, by which it defines a narrow market which is not sufficiently covered, and makes efforts to conquer it through strong promotional activities. Such a strategy is possible in the markets dominated by major players, and the venture can emerge as a modifier or a provider of specific solutions for various intended uses of the existing solutions, when it comes to products/services. The strategy of gradual conquering of the market is intended for those solutions that have a potential of becoming universal solutions in certain fields. Such products require substantial investments, powerful promotional activities, firm distribution channels and post-sales support. In such cases, researchers often leave universities in order to commit to the company's growth and development.

Due to research potential, university venture can be organized, financed and managed in line with the interest of the key players when it comes to development and production of certain products. Namely, a strategy which can be offered to key players in the market, essentially involves product modification or product improvement, leading to data collection with the purpose of learning about consumers' preferences and habits, which is of great importance in the circumstances of intensive changes and large investments in research and development.

Once a strategy has been chosen, it poses a serious task to a university venture, requiring that it defines the so-called control points by which it establishes the appropriateness of the chosen strategy by controlling the defined objectives. In other words, the strategy must have defined goals (points) that are to be met at certain time intervals. In this way, it is possible to measure the success level of the selected methodology. If the defined objectives are fulfilled or even surpassed, the selected strategy should not be changed, but only improved and adjusted to the market's needs. On the other hand, if the strategy is lagging behind the control points, this may imply that the selected strategy is not appropriate for the current situation in the market. This leads to the conclusion that control points are necessary to preserve resources and to ensure their efficient use. Without such control points, a recklessly defined strategy may cause wasting the resources, both human and financial, and it would fail to meet the expected outcomes, which will affect the further development of the university venture. Such control points may have short-term deadlines, extending up to one year, but they can also be long-term, extending up to five or ten years.

2.5 *Team Creation*

During the initial stage of developing a university venture, one has to think of creating a team for research commercialization, the team that will be capable of implementing strategic decisions related to market positioning. The research results have shown that researchers are not good entrepreneurs, because they have little success in implementing even those strategies which are well defined and in line with the market needs. For this reason, it is crucial to create a good team at the beginning of the university-venture development, particularly if there is a decision to put external experts on the team, the external experts to include people experienced in company management.

The main barrier to achieving the success of a university venture is related to the participant's willingness to share responsibilities in commercializing research. Researchers, who have invested huge efforts in developing a product/service, are seldom ready to share a part of the responsibility with experienced experts, holding the view that they themselves are the only ones deserving of credit for the product/service. For this reason, projects often fail because researchers are often not experienced enough in securing the market and positioning within it. Team creation must be aligned to the dynamics of the environment in which a university venture operates, and, for this reason, it is important to reduce paper work and ensure that there are flexible relations allowing for fast exchange of information and delegation of decision making and taking responsibilities.

Creating a knowledge organization is the task of a university venture intended to track, develop and apply knowledge in order to conquer a market and to build on it. The most serious mistake one could make is to conquer the market and then direct the resources towards defending the position there. Such an organization allows for creating a system for constant evaluation and control of not only the processes, but also the market performance, the activities of competitors and the achieved financial results.

3 Filling the Gap in Financing University Ventures

After making a decision to launch a university venture, it is important to consider the financial matters, or, in other words, it is imperative to secure the capital for commercializing research. Since this funding will be used for a business venture that is still in its early stages and whose market is yet to be confirmed, financial sources are normally limited. Traditional financial sources, including the banking sector or financial market, are unavailable, so that the university venture must rely on its own financial sources, loans or support from friends and family, and on contacts with specific financial sources, including business angels and venture-capital funds (Cumming and Johan 2009). Depending on the products/services being developed by a university venture, key players can be contacted who are willing to take on substantial risk in order to secure adequate financial resources. Every venture faces limited financial resources in its initial stages, but one can also work under the peaking theory, i.e. that the best source of financing includes one's own resources, followed by loans, and finally, the last source to be sought should be equity investments.

The peaking-order theory implies that every entrepreneur can create his/her own acceptability order when it comes to various finance sources. In other words, every entrepreneur wants to finance his/her own venture with her/his own sources in order to achieve the complete success by themselves, thus, limiting the probability of a potential failure. If personal resources are not enough, the entrepreneur is willing to apply for a bank loan, accepting a given interest rate and providing collateral, with a potential loss of private or business property, depending on the company structure and organization. The entrepreneurs may opt for taking out a loan in those cases when it is expected to achieve a performance that will enable them to pay back the loan, without compromising the growth and development of the company.

Finally, according to this theory, entrepreneurs opt for equity investment which implies a loss of a part of, or the total, management control of the company. This is due to the fact that investors take over a share of the ownership and, in this way, control the business performance of the company. Certainly, a high price of equity funding results from the high risk undertaken by the investors because, in the case of failure, they are unable to recoup their resources on account of the company's property, or the property of other owners. In general, start-ups and other companies in early stages of development are suitable for this investment scheme, since banks and financial markets are reluctant when it comes to investing in companies which haven't established their position in the market.

Regarding university ventures, they are regarded as highly risky because they are established and run by individuals with no, or with little, experience in creating and running a business. Such characteristics aggravate their own position when trying to provide finances because their own funds are limited. For this reason, it is necessary to attract financial intermediaries ready to assume the high-risk level and provide capital to the university ventures (Metric 2007). Such sources include: business angels, venture capital funds and crowd sourcing financing.

It is essential to strengthen the venture's presence in the capital industry through both formal and informal channels (Gompers and Lerner 2004), in order to create foundations for establishing investment potential for the venture. In terms of informal channels, key players are identified as wealthy individuals ready to invest their own money at various stages of the business' development. They are often former executives investing in familiar business fields, using contacts and experience in building up the business project. They are selecting a non-listed company with adequate growth opportunities, acquire ownership (not a majority position), work with the acquired company and later sell shares through an IPO on the financial market. Relying on the existing research, business angels are willing to invest €50,000–500,000 per qualified university-venture project.

However, it is even more important to know that business angels do not bring only money (for which they get a share in the ownership), but they are also willing to help the owners with development, expansion and market positioning. Evidently, they are motivated to act this way in order to increase the value of the venture, so that they could recoup their investment and enjoy a satisfactory profit rate. Once they have invested in a venture, business angels demand that they should have financial control over their resources, by imposing strict rules as to what the money they invested could be used for. Certainly, this capital will not be used for satisfying the personal needs of other owners, but will be used to improve products/services, to organize distribution channels, and to take care of the promotional activities. If they invest their own money, business angels will rely on their own experience and their contacts to make the venture visible and recognizable on the market. This can be even more important than the money they invest, because their contacts and their market experience can increase the venture's chances for success.

Business angels are quite experienced market people with deep knowledge and limited time, thus, during the negotiating process, venture owners need to take an open and transparent approach, with nothing to hide. Such an approach is needed because the relationship between a business angel and venture owner will be long lasting and even on a daily basis. As they invest their own money, business angels often base their decisions on their own impressions, hence, they often prefer: "that it is better to invest in a poor project with a good management, than in a good project, with a poor management."

Institutional venture-capital firms (Gompers and Lerner 2001) represent professional funds working as intermediaries between capital providers (various financial institutions) and capital searchers (unquoted companies). Their primary task is to find interesting business projects with growth opportunities and then to convince financial institutions to invest money. Financial institutions are pension funds, insurance companies, wealthy families etc. The organizational forms of the institutional venture capital can be varied, depending on the ownership structure:

- independent limited partnerships, in which the venture capital firm serves as the general partner, raising capital from limited partners such as institutional investors (pension funds, insurance companies and banks);
- captive venture-capital firms which are mainly funded by the internal resources of a parent organization—a bank or an insurance company, or by non-financial institutions;
- governmental venture-capital organizations which are financed and controlled by government institutions.

As the third group within the venture-capital industry, corporate venture capital is mentioned. It is defined as equity or equity-linked investments in young privately held companies, where the investor is a financial intermediary of a non-financial corporation. The corporate venture capital is organized by a large corporation in order to develop a specific product or technology that can be related to the current business or it can be totally new, providing new sources of profit.

The main difference between non-formal and formal ventures is in the sum of the available resources and the resources invested in a project. Specifically, thanks to their being managed by a group of professionals and due to the fact that venture-capital funds collect resources in a number of ways, they can collect between dozens to hundreds of millions of euros, depending on the size of the industry. Surely, the biggest venture-capital funds can be found in the USA, which is, at the same time, the state where this kind of industry is the most developed worldwide. In addition, the money invested in individual projects start with one and go up to dozens of millions of dollars, which is much more than the funding that can be expected from business angels.

Another difference between these two kinds of investments lies in the stage of investment. Venture-capital funds make investments during the stages that follow the initial stages in the life of a university venture. In other words, such funds are sought by the companies that have already experienced a certain success in the market, so now they seek to find a stronger support in order to improve the product, to become distinguished in the market and to enhance the distribution channels. In this way, venture capital funds finance the companies whose concept is already recognized and that have already secured a certain position within the market. However, although they make selections in this manner, it is still possible that commercial success will not be achieved in all cases. Research results have shown that that one out of five projects survives and justifies the resources coming from a venture-capital fund.

The last resort in securing financial support for a university venture can be the so-called crowdfunding, which is based on a software platform within which it is possible to present projects and collect resources. Financing like this is based on the tradition of organizing investment bashes during which projects are presented to the interested investors. More recently they are functioning within online systems for project application and analysis, and such systems can be private or public or it can take the form of a public-private partnership. Another form of platform organization is through private initiative, which will assume profits from collecting registration fee from projects, registered investors and commissions coming from the projects with finance secured through platform. Alternatively, the platform can be state-owned, in the case when the state, through one of its institutions, develops the platform as a means to support the development of alternative sources of financing. Finally, creating such platform can be a joint investment of the state and private sector, with clearly defined ownership rights and management rules for both sides.

Regardless of the organizational structure, crowdfunding is an interesting source of project financing, especially for projects that do not require substantial initial investments, and as such they are disqualified by business angels and venture-capital funds. The idea behind crowdfunding is to secure a large number of small investors willing to direct their own money to the projects registered on the platform. In this way, it is possible to create a pool of potential investors who can, by investing smaller portions of money, become active players in the project's growth and its development. Certainly, depending on the level of their investment, these investors can expect a share in the company, as well as they can expect to be able to participate in company's management.

This process begins with the registration of projects on the platform, where the platform guarantees confidentiality of data and clearly warns investors against data abuse, which is punishable by law. During the registration process, the project owner describes the project idea and provides a general overview, stating why the project has potential for commercialisation. Parallel to the registration by the project's owner, investors get registered on the platform, accepting the conditions of the platform, in particular those related to data protection. Some platforms require that users make a down payment, which may not be retrieved in cases of data abuse. Upon the registration, investors are granted the right to access the project archive, containing the main data about the project. Should they become interested in a project, investors get in touch with the project owner, upon which negotiations can take place, in line with a mutual agreement between the project owner and investors? If they agree to the terms of the co-operation, the negotiations are continued outside the platform, i.e. via direct contact between the capital provider and venture owner.

The second possible solution on some platforms is related to the situation when all the processes, including analyses, negotiations and investments take place on the platform. In this way, the owners of the platform take full control of the process, and they avoid the situation when money cannot be collected from the investors. This type of crowdfunding platform organization is also present in the situations when a large number of small investors invest their money, and in return, get a small percentage of ownership. In these circumstances, it is pointless to try to organize board meetings, as no one has greater influence over the company than the owners themselves.

Regarding crowdfunding organizations, there is evidence of some typical ways of inclusion of investors. Firstly, investors perform as philanthropists, investing their money to support an idea or a project, not expecting anything in return. Secondly, investors make investments in projects, and in return they have the possibility of buying the project/service before anyone else, or at a price below the market price. Thirdly, investors make loans for projects, but at slightly higher interest rates than those in the banking sector, and this is because of the higher risk that is associated with the venture. Finally, the most frequent way for investors to perform is to invest their money, expecting an equity stake in the projects, which become enterprises and thus attract the attention of the public as a possible additional source of financing.

Crowdfunding can be a very interesting source of financing university ventures, particularly if we consider the fact that the total sum investors provide depends entirely on their free will and risk aversion. In other words, if we consider the way universities are organized, it seems fairly easy to upload a platform to the academic network, and in this way create a starting point for development of a specific source of financing for university ventures in their initial stages. Every university has income coming from various commercial activities, and such income can become capital to be invested in creating the platform, as well as it can be invested in potentially promising ideas. It is highly recommended to hire external experts experienced in equity investment, who will take part in the selection of the promising ideas in order to avoid conflict of interest in universities. In return, external experts can be granted free access to the platform, i.e. the ideas coming from the university. In fact, the development of the platform will be based on the cooperation between public and private sectors, in line with recommendations based on numerous research results that show that the state is acutely inefficient when it comes to making decisions related to equity financing.

The advantage to the university crowd financing lies in fast information exchange, familiarity with the people writing project proposals, transparency and the potentially large number of investors from the pool of teaching staff. On the other hand, the academic environment tends to be slightly closed, with researchers unwilling to share their research results. However, in the circumstances of a system built in a way to motivate researchers to become entrepreneurs, such problems will no longer exist. On the contrary, researchers will be interested in exchanging experiences, as well as in working together on enhancement of products/services.

3.1 Precondition for University Venture Financing

The main issue regarding university ventures as business enterprises, is in developing an organizational system that will motivate specific financial resources to turn towards university enterprises as potential profitable ventures. The state should work on creating appropriate preconditions, establishing more financial tools for financing the initial stages of university business ventures.

There are various efforts aimed (Landstrom 2007) at creating such systems, but one that stands out is the US SBA program, which is based on co-financing, and whose role is to motivate venture capital to play the role of a financing resource. After a number of efforts, the EU accepted this concept and developed programs based on the co-financing scheme between public and private resources. In other words, the state provides financing resources during early stages in the life cycle of an enterprise, and after that it opens a call for private investors to apply for resources. After the resources have been awarded, the state steps away from managing the resources, but retains a control role and its right to retrieve the resources if any incident of abuse occurs. The main justification for such practice is the fact that private capital is far more efficient in selecting and managing the company into which the money is invested (Lerner 2010). The supporting programs, defining the state as an active investor, have proven to be highly non-transparent, inefficient and lacking in any instrument to measure the profitability of selected projects. On the other hand, private investors carefully select projects because the success of one project recommends them for the next one, and their reputation in the world of venture-capital investments becomes stronger.

When it comes to the role of the state, there are several ways in which the state can influence the creation of a venture-capital industry. Firstly, in the case of business angels, the state can use its tax policy to motivate rich individuals to invest a part of their capital into new companies. States do that, but they should pay special attention to the individuals who are willing to invest in university ventures. Such individuals should be entitled to significant tax deductions because in such circumstances they are likely to assume higher risks and make conditions for creating university ventures. At the same time, they will help increasing the level of entrepreneurship at universities both among students and researchers. A success story, or at least the awareness of the possibility to get their projects financed by business angels, will raise the awareness of staff and students, resulting in a better economic environment in the region, as well as in the country.

The EU made a step forward and in 2012 promoted the European Angel Fund, based on principles of co-investment with private sector. In particular, if they show experience in financing enterprises in their initial stages, individuals are qualified to be listed as business angels, eligible for grants from the fund. In this way, the EU develops not only tax-deduction mechanisms, which exist at the level of individual states, but an additional mechanism as well, in order to motivate individuals to become business angels.

In the case of formal venture capital, i.e. venture-capital funds, the support measures include tax deductions, loans with lower interest rates, insurance for investments, co-investment and business incubators or accelerators. Countries find various different paths and rely on various experiences in their efforts to spur the venture-capital industry. One group of countries works on developing an appropriate tax system to motivate the flow of investments into venture-capital funds, but these states encourage creation of new funds as well. In particular, countries provide deductions for investing in funds, but they also provide tax deductions for income coming from venture-capital-fund business activities. The other group of countries works on establishing public-private partnership through securing additional resources for investments during early stages of development of enterprises. The difference between these two approaches is related to the way in which the countries "subsidize" private venture-capital funds.

One group of countries entering the partnership does not demand a minimum price but they act as active investors, sharing with private investors not only profits but risks as well. Other countries secure resources under conditions that require that minimum resources should be returned in the form of interest. In this way, the state imposes additional pressure on private investors when they select appropriate projects. The third group of countries offers insurance to compensate investors in the case of losses caused by projects that fail. More specifically, the state guarantees a certain percentage of the resources intended for venture-capital investment, and, in the case of failure, it pays out this percentage to the fund. Such support proves to be ineffective and research results suggest that this is the case due to the relative security coming from the insurance, and, as such it affects the selection of the projects. The fourth group of countries includes those where public funds individually select projects and implement them.

The last support group includes those countries that create business incubators that provide not financial but logistic support to companies in the early stages of development. In such programs, enterprises get facilities and infrastructure at costs below market, but only for a limited time. The incubators also organize free business-management courses.

4 Strategies in University-Venture Growth

If university ventures show success in solving problems typical of early stages of their development and achieve success in research commercialisation, there is the question of what to do next. Should the university venture proceed with expansion and growth, or should it be satisfied with its current position? Although this question may seem simple, in reality, it is as difficult as considering whether to start a university question or not. Getting the products/services commercialised and recognized on the market is one thing, whereas the matter of expansion is something else, thus, one must think of the risks associated with this idea. The decision will be made by the owner, but, in the case when the venture is financed by various sources of financing, the decision will be based on a compromise with the investors. If we assume that it is the researcher who makes this decision, then the main problem to it provide the needed time and resources. If they want to expand their businesses, researchers-entrepreneurs must dedicate more time to their business, which may have an adverse effect on their academic duties. It is very hard to find the right balance between the time needed for regular teaching activities and the time needed for expanding the business. For this reason, some professors give up their teaching career and commit to commercialisation of their research.

In the situations where investors are present, it is certain that they will put additional pressure on researchers in order to convince them to expand the business, because investors are interested in gaining higher profits. Investors can suggest that the business should expand, excluding active participation of the researcher, provided that the researcher should be given a smaller portion in the ownership structure, and, accordingly, a certain percentage of the profits.

When we talk about the growth of a university venture, we must consider that there are several alternatives to its future plans regarding growth and the market. More specifically, there are decisions to be made as to whether to expand the existing market, to develop a new market or to work on continuous product development. The expansion of the existing market is the simplest strategy because of the clearly defined market that need to be conquered piece by piece, in line with the appropriate strategies. Depending on the condition of the industry, conquering the market may require that different strategies should be used. It is certain that fast-growing industries call for intensive communication with consumers and constant improvement of the services offered to them. Additionally, when there is a large number of smaller players in the industry, it is necessary to work on making the products/service recognisable through innovative marketing campaigns and strengthening of support networks.

Market development assumes potential expansion of its boundaries in relation to the original ones. Surely, depending on where the products are used, markets can be local, regional, national or global. If a product has local or regional character, the expansion strategies are limited, whereas the global character offers possibilities for discovering new markets and generating revenues within them. Nevertheless, performance in new markets calls for special preparations because conditions vary in different markets, therefore it is impossible to use the same strategy in all markets. Expanding the market at a global level assumes careful selection of the appropriate performance strategy, including possible product modifications.

A university venture can enter the market either independently or in partnership with local entrepreneurs. If it opts for the latter, it can so implement it via franchising or licencing. A franchise is based on selling the rights to use the products, advertisements, production organization and quality assurance once the business venture has been settled. A university venture can sell the franchise and work only on monitoring the sustainability of the products/services. The negotiation process itself can last long and can take a lot of effort because many things need to be considered in order to protect both parties. This process protects the franchise owner in all instances of abuse or misuse of the product/service, while, at the same time, unification of the product/service is guaranteed. On the other hand, licencing is a means by which a university venture authorizes the second party to use its intellectual property, allowing them to organize production, sales, marketing and post-sale services.

Regardless of the situation, university ventures can use several models in expanding their business activities. Firstly, university ventures can opt for independent performance in market expansion, in which case it may opt for several strategies:

- direct "attack" against the key players in the market, which is a very unlikely strategy because it requires substantial resources and experience in playing the market game, given the fact that this strategy implies winning traditional consumers of the existing products and services;
- "indirect attack" against the key players in subsidiary markets, where competitors are weak and unprepared for the game; thus, this strategy is applied by developing specific products that meet the market demands;
- "guerrilla attack" against the key players in the market by a special strategy aimed at covering small portions of the market by quick moves back and forth, without engaging significant resources.

If it concludes that it is not possible to opt for independent performance in the market, the university venture can decide to cooperate with other companies that are operating there. The possible modules include:

- Strategic partnership with key players. This partnership can be implemented through agreements on complementing the existing supply of the leading

competitors by a product/service developed by a university venture. Also, the university venture may agree to authorize leading companies to sell its product in the market.
- University venture may opt for merger with leading companies in the industry, to secure the market and additional resources, but, in this process, the venture loses part of its independence in making decisions.

5 Successful Stories in University Ventures

It is hard to make a clear classification of university ventures with respect to the way they are organized because of various legislative solutions, as well as the possibility to position the venture within a university or to affiliate it with a university as a means of commercialising scientific research. The issues that need to be considered when organizing a university venture are the following:

- Who is in charge of organizing a venture? Observing university ventures worldwide, it can be noted that they can be organized by professors, researchers or by students who want to turn their research results into a commercial success. In some cases, university ventures result from the cooperation between professors and students in doctoral research programs, where the professor takes the role of a mentor in the initial stages of research; whereas in the case of a success, the professor gets a position on the managing board of the company.
- Who provides financing to a university venture? Financing of a university venture can be regulated by law or a book of rules and procedures issued by a scientific research institution. Some institutions allow the researchers to seek funds for commercialisation from professional investors (business angels, venture capital funds etc.), while some other institutions create their own sources of financing by forming venture-capital funds using their own resources, as well as the resources resulting from successful projects of research commercialisation.
- How does one organize the exploitation of university resources? This is a key question when it comes to commercialising scientific research because it deals with compensation provided for using the university resources. Some universities set fixed tariffs for these types of compensation, others require ownership of the product or a service developed by the venture, while there are some universities that demand to have shares in the company allowing them to collect part of the profits, as well.

Regardless of the organization scheme, university ventures are seen as a solid basis for strengthening the economy and providing employment to the work force, in particular to the young scientists, who are most vulnerable nowadays. For this reason, all developed countries work on programs related to providing support to university ventures. As a result, millions of dollars/euros are invested in development of specific support programs that facilitate financing the commercialisation of scientific research. Thanks to such efforts, there is a large number of university ventures that have found their own place on the market, where they perform with excellent financial results. In the following sequence, there are listed some success stories, chosen by the authors, without any intention to neglect others or to undermine their success:

Wave—the company originating from a research related to wireless charging of buses in urban areas. The company was founded at Utah State University, and so far it has had several commercial engagements. The main advantage to technology developed by Wave lies in the possibility of charging buses on their regular routes, by placing wireless chargers at the bus stops. In this way it is possible to charge buses all day long, making the ride economical, safe and profitable for urban-service providers.

Gradiant—the company founded in 2011 at MIT as a result of research dedicated to energy and environmental protection. The objective of the research and subsequent commercialisation was to offer a new way of water treatment and desalination and to produce fresh water from high concentration brines.

EyeNetra—the company founded in 2012, at MIT, which developed a completely new way of detecting sight problems, by using a cell phone application.

Slips Technologies—Harvard University has spun off this venture, using research at Harvard Wyss Institute for Biologically Inspired Engineering, in 2014. The research is related to the development of resistant material for any kind of liquid and solid. SLIPS (Slippery Liquid-Infused Porous Surfaces) are a pioneering and award-winning set of technologies that transform the surface of a solid material into a microscopically thin and ultra-smooth (friction-free) immobilized "sea" of lubricant.

Circassia—the company dealing with development of technologies in the field of medicine, specializing in new forms of allergy treatment. It emerged thanks to financial and organizational support from the Imperial Innovations investment fund which invested 115 million pounds collected from private investors, who were provided with a confirmation that the new technology shows positive results in allergy treatment. In March 2014, the company decided to secure additional capital on the London Stock Exchange and got listed on the exchange, which helped to obtain 200-million pounds and achieve 581-million pounds of market capitalization.

Imperial Innovations is a fund resulting from the need to foster the commercialisation of scientific research at Imperial College, University of Oxford, University of Cambridge and University College London through:

- leading the formation of new companies,
- providing operational expertise and facilities in the early stages,
- providing substantial investment capability and facilitating co-investment to accelerate development,
- helping to recruit high-calibre executive management teams and board members in portfolio companies.

1248—the company founded at University of Cambridge, representing a new model of connecting products and services to consumers by the Internet. In particular, the company offers a new way to control the functioning of products and services after sales, without the need to develop the infrastructure necessary to monitor the progress individually. 1248 resulted from a need to monitor the progress of products and services delivered for other production processes in a simple way, and, as such, they constitute an important factor for the success of a company. 1248 developed a platform by which it is possible to monitor the status and to predict the future state of the delivered technology, as well as the need to intervene, when necessary.

Alesi Surgical—the company which emerged as a result of the scientific research activity at Welsh Institute for Minimal Access Therapy at Cardiff University in 2009. The idea used as a basis for founding the company was related to the removal of hazardous vapour, known, also, as surgical smoke, produced by medical instrument during surgical interventions. The name of the product is Ultravision, and at present, this company distributes its products in 25 countries worldwide.

Clyde Biosciences—this company, founded at the University of Glasgow, within the College of Medical Veterinary and Life Science, is committed to the development of new technology for assessing drug toxicity. This methodology is used in the early stages of drug development for the purpose of detecting drug efficiency and toxicity through measuring drug- induced changes in Action Potential. After initial success in applying new methodology for developing medicament, this company started the research related to the development of a cure for Parkinson's disease.

Spirogen—this company, in the field of biotechnology, was founded in 2001 at the University College London. It is focused on the development of antibody-drug technology that targets cancer tumours. Thanks to the excellent results it has achieved, it was sold to a pharmaceutical company in 2013 for 273 million pounds.

NuCana Biomed—this company, dealing with research related to development of anti-cancer drugs, was founded in 2008 by the Edinburgh University. The developments fostered by this company bypass the major mechanisms associated with anti-cancer drugs.

Lycotec—this company was founded by Cambridge University, and it is famous for its anti-aging chocolate bars, sprout and clinical nutrition, nutra-cosmetics and nutraceuticals. Lycotec focuses on the development and commercialisation of its proprietary bio-medical technologies for pharmaceutical, nutraceutical, functional food, cosmetic and personalised nutrition industries.

6 Conclusions

University ventures are an important task for state regulative and policy makers in today's intensive economically and socially competitive world. Policy makers need to develop such supporting programs that will create business thinking among researchers, making them more involved in the real world, with active competition and profit-making possibilities.

Creation of entrepreneurial universities is the first step, which will change and challenge traditional ways of thinking among majority of researchers, not already familiar with, and not already approving, a business-research orientation. Current intensive competition among companies and states in technologies and innovations produces the necessity of tight cooperation between education institutions and business entities.

The universities need to work on deeper involvement in the active business community, which is crucial for understanding the surroundings for conducting business activities. The involvement assumes collecting information about market conditions, economic conditions on the macro level (GDP, employment, inflation, foreign trade etc.), industry-specific information and possibilities to expand abroad with developed products and services. This information is necessary for creating university ventures on a commercial basis, oriented to profit by achieving a certain market share and fulfilling the defined customers' needs.

The university research staff needs to acquire a certain level of knowledge about market analysis, venture managing and team creation and management. These are the major challenges for university ventures, led by researchers not familiar with business models, procedures and strategies, and future efforts need to be addressed to create and develop such skills. Specific problems that will be challenges for university ventures are a scarcity of financial products ready to involve themselves in highly-risky business ventures. Standard financial products, such as bank credit and financial-markets securities, are not suitable for university ventures, so new ones need to be developed.

Investors, not risk averse, ready to invest their money, time, knowledge and experience, are the ones needed for university ventures because they meet the above mentioned criteria. Such financial products are possible if business angles, venture capital funds and crowdfunding platforms are present. Business angels, venture-capital funds and crowdfunding platforms constitute the equity investors looking for prospective companies in their initial phases. They are investing money, experience, contacts and time working along with venture owner, for equity stakes that are not the majority but with strong financial control. Their interest is venture growth, by which their equity stakes grow also and become more valuable at the market. The state needs to create an environment for equity investors because they are willing to take risk and finance university ventures in their initial phases.

The university ventures' role in dynamic economic surroundings is quite important if a country would like to preserve or increase its competitiveness on the regional and global markets. Researchers need to leave their secure places, within the university surrounding and state project financing, and develop the necessary skills for market competition and profit earning. Such movement towards the market will open universities for more intensive contacts with business communities, which will result in curriculum and program adjustment according to business needs. Also, university ventures will open up the possibility for students to work with professors and researchers on concrete business projects, helping them to understand and acquire experience in the real world, outside the classroom.

References

Abreu M, Grinevich V (2013) The nature of academic entrepreneurship in the UK: widening the focus on entrepreneurial activities. Res Policy 42(2):408–422

Acs ZJ, Braunerhjelm P, Audretsch DB, Carlsson B (2009) The knowledge spillover theory of entrepreneurship. Small Bus Econ 32:15–30

Allen S, Link A, Rosenbaum D (2007) Entrepreneurship and human capital: evidence of patenting activity from the academic sector. Entrep Theory Pract 31(4):937–951

Ambos TC, Makela K, Birkinshaw J, D'Este P (2008) When does university research get commercialized? Creating ambidexterity in research institutions. J Manage Stud 45(8):1424–1447

Audretsch DB, Lehmann E, Warning S (2005) University spillovers and new firm location. Res Policy 34(7):113–1122

Audretsch D, Lehrmann E (2005) Does the knowledge spillover theory of entrepreneurship hold for regions? Res Policy 34(8):1191–1202

Baldini N, Grimaldi R, Sobrero M (2006) Institutional changes and the commercialization of academic knowledge: a study of Italian universities' patenting activities between 1965 and 2002. Res Policy 35(4):518–532

Baumol WJ (2005) Education for innovation: entrepreneurial breakthroughs vs corporate incremental improvements. In: Jaffe AB, Lerner J, Stern S (eds) Innovation policy and the economy. The MIT Press, Cambridge, pp 33–56

Benjamin G, Margulis J (2001) The angel investor's handbook. Bloomberg Press, Princeton

Bercovitz J, Feldman M (2008) Academic entrepreneurs: organizational change at the individual level. Organ Sci 19(1):69–89

Central and Eastern Europe (2009) EVCA special paper, EVCA, PREP Analytics, July 2009

Chen H, Gompers P, Kover A, Lerner J (2009) Buy local? The geography of successful and unsuccessful venture capital expansion. Harvard Business School, Harvard

Colombo MG, D'Adda D, Piva E (2010) The contribution of university research to the growth of academic start-ups: an empirical analysis. J Technol Transfer 35:113–140

Colyvas JA, Powell WW (2006) Roads to institutionalization: the remaking of boundaries between public and private science. Res Organ Behav 27:305–353

Cumming D, Johan S (2009) Venture capital and private equity contracting. Elsevier, Burlington

Dearlove J (2002) A continuing role of academics: the governance of UK universities in the post-Dearing era. High Educ Q 53(3):257–275

Debackere K, Veugelers R (2005) The role of academic technology transfer organizations in improving industry science links. Res Policy 34(3):321–342

Ding WW, Choi E (2011) Divergent paths to commercial science: a comparison of scientists' founding and advising activities. Res Policy 40:69–80

Djokovic D, Souitaris V (2008) Spinouts from academic institutions: a literature review with suggestions for further research. J Technol Transfer 33(3):225–247

Dohse D, Walter S (2012) Knowledge context and entrepreneurial intentions among students. Small Bus Econ 39(4):877–895
Etzkowitz H (2004) The evolution of the entrepreneurial university. Int J Technol Glob 1(1):64–77
Eun J, Lee K, Wu G (2005) Explaining the university-run enterprises in China: a theoretical framework for university–industry relationship in developing countries and its application to China. Res Policy 35(9):1329–1346
EVCA White Venture Capital Paper (2010) Closing gap and moving up gear: the next stage of Venture capital's evolution in Europe, EVCA, Brussels. 2 Mar 2010
Falck O, Heblich S, Luedemann E (2012) Identity and entrepreneurship: do school peers shape entrepreneurial intentions? Small Bus Econ 39(1):39–59
Gilson J (2003) Engineering a venture capital market: lessons from American experience. Stanf Law Rev 55(4):1067–1103
Gladstone D, Gladstone L (2004) Venture capital investing. Ft Prentice Hall, New Jersey
Goethner M, Obschonka M, Silbereisen R, Cantner U (2012) Scientists' transition to academic entrepreneurship: economic and psychological determinants. J Econ Psychol 33:628–641
Goktepe-Hulten D, Mahagaonkar P (2010) Inventing and patenting activities of scientists: in the expectation of money or reputation? J Technol Transfer 35:401–423
Gompers P, Lerner J (2004) The venture capital cycle. The MIT Press, London
Gompers P, Lerner J (2001) The venture capital revolution. J Econ Perspect 15(2):145–168
Guerrero M, Urbano D (2012) The development of an entrepreneurial university. J Technol Transfer 37(1):43–74
Haas MR, Park S (2010) To share or not to share? Professional norms, reference groups, and information withholding among life scientists. Organ Sci 21(4):873–891
Henkel M (2007) Can academic autonomy survive in the knowledge society? A perspective from Britain. High Educ Res Dev 26(1):87–99
Hogan T, Hutson E (2005) What factors determine the use of venture capital? Evidence from Irish software sector. Venture Cap 7(3):259–283
Horlings E, Gurney T (2013) Search strategies along the academic lifecycle. Scientometrics 94:1137–1160
http://www.imperialinnovations.co.uk/about/. Accessed 15 Feb 2017
http://slipstechnologies.com/about-slips.php. Accessed 15 Feb 2017
http://www.waveipt.com/#!/projects. Accessed 15 Feb 2017
http://1248.io/about.php. Accessed 15 Feb 2017
http://www.circassia.com/technology/. Accessed 15 Feb 2017
Jain S, George G, Maltarich M (2009) Academics or entrepreneurs? Investigating role identity modification of university scientists involved in commercialization activity. Res Policy 38:922–935
Karlsson T, Wigren C (2012) Start-ups among university employees: the influence of legitimacy, human capital and social capital. J Technol Transfer 37(3):297–312
Krabel S, Mueller P (2009) What drives scientists to start their own company? An empirical investigation of Max Planck Society scientists. Res Policy 38(6):947–956
Lam A (2007) Knowledge networks and careers: academic scientists in industry–university links. J Manage Stud 44(6):993–1016
Lam A (2010) From 'ivory tower traditionalists' to 'entrepreneurial scientists'? Academic scientists in fuzzy university–industry boundaries. Soc Stud Sci 40(2):307–340
Landstrom H (2007) Handbook of research on venture capital. Edward Elgar Publishing Limited, Cheltenham
LeBreton JM, Senter JL (2008) Answers to twenty questions about interrater reliability and interrater agreement. Organ Res Methods 11:815–852
Lee L, Wong PK, Foo MD, Leung A (2011) Entrepreneurial intentions: the influence of organizational and individual factors. J Bus Ventur 26:124–136
Lerner J (2010) Boulevard of broken dreams. Why public efforts to boost entrepreneurship and venture capital have failed and to do about it. Princeton University Press, Princeton

Link A, Siegel D, Bozeman B (2007) An empirical analysis of the propensity of academics to engage in informal university technology transfer. Ind Corp Change 16(4):641–655

Link AN, Ruhm CJ (2011) Public knowledge, private knowledge: the intellectual capital of entrepreneurs. Small Bus Econ 36:1–14

Lockett A, Wright M (2005) Resources, capabilities, risk capital and the creation of spin-out companies. Res Policy 34(7):1043–1057

Markman GD, Siegel DS, Wright M (2008) Research and technology commercialization. J Manage Stud 45(8):1401–1423

Martinelli A, Meyer M, von Tunzelmann N (2008) Becoming an entrepreneurial university? A case study of knowledge exchange relationship and faculty attitudes in a medium-sized, research-oriented university. J Technol Transfer 33(2):259–283

Metric A (2007) Venture capital and finance of innovation. Couriere/Westford, New Jersey

Mosey S, Noke H, Binks M (2012) The influence of human and social capital upon the entrepreneurial intentions and destinations of academics. Technol Anal Strateg Manag 24 (9):893–910

Murray F, Graham L (2007) Buying science & selling science: gender stratification in commercial science. Ind Corpor Change Spec Issue Technol Transf 16(4):657–689

Nicolaou N, Birley S (2003) Academic networks in a trichotomous categorization of university spinouts. J Bus Ventur 18:333–359

Nosella A, Grimaldi R (2009) University-level mechanisms supporting the creation of new companies: an analysis of Italian academic spin-offs. Technol Anal Strateg Manag 21(6): 679–698

O'Shea RP, Allen TJ, Chevalier A, Roche F (2005) Entrepreneurial orientation, technology transfer and spinoff performance of U.S. universities. Res Policy 34(7):994–1009

Ping C (1980) Industry and the universities: developing cooperative research relationships in the national interest. National Commission on Research, Washington

Powers J (2003) Commercializing academic research: resource effects on performance of university technology transfer. J High Educ 74(1):26–50

Prodan I, Drnovsek M (2010) Conceptualizing academic-entrepreneurial intentions: an empirical test. Technovation 30:332–347

Rasmussen E, Borch OJ (2010) University capabilities in facilitating entrepreneurship: a longitudinal study of spin-off ventures at mid-range universities. Res Policy 39(5):602–612

Rasmussen E, Moen O, Gulbrandsen M (2006) Initiatives to promote commercialization of university knowledge. Technovation 26(4):518–533

Rothaermel F, Agung S, Jiang L (2007) University entrepreneurship: a taxonomy of the literature. Ind Corp Change 16(4):691–791

Schiller R (2005) Irrational exuberance. Currency Doubleday, New York

Schneider B, Ehrhart MG, Macey WH (2013) Organizational climate and culture. Annu Rev Psychol 64:361–388

Siegel DS, Veugelers R, Wright M (2007) Technology transfer offices and commercialization of university intellectual property: performance and policy implications. Oxf Rev Econ Policy 23 (4):640–660

Siegel DS, Waldman DA, Atwater LE, Link AN (2004) Toward a model of the effective transfer of scientific knowledge from academicians to practitioners: qualitative evidence from the commercialization of university technologies. J Eng Tech Manage 21:115–142

Siegel DS, Waldman D, Link AN (2003) Assessing the impact of organizational practices on the relative productivity of university technology transfer offices: an exploratory study. Res Policy 32(1):27–48

Souitaris V, Zerbinati S, Al-Laham A (2007) Do entrepreneurship programmes raise entrepreneurial intention of science and engineering students? The effect of learning, inspiration and resources. J Bus Ventur 22:566–592

Staebler M (2009) Description of the small business investment company program participation by funds using debentures. Pepperlaw, Detroit

Stuart TE, Ding WW (2006) When do scientists become entrepreneurs? The social structural antecedents of commercial activity in the academic life sciences. Am J Sociol 112(1):97–144
Tijssen R (2006) Universities and industrially relevant science: towards measurement models and indicators of entrepreneurial orientation. Res Policy 35(10):1569–1585
TyVan Looy B, Landoni P, Callaert J, van Pottelsberghe B, Sapsalis E, Debackere K (2011) Entrepreneurial effectiveness of European universities: an empirical assessment of antecedents and trade-offs. Res Policy 40(4):553–564
Walter S, Parboteeah K, Walter A (2011) University departments and self-employment intentions of business students: a cross-level analysis. Entrepreneurship Theory and Practice 37(2): 175–200
West J (2008) Commercializing open science: deep space communications as the lead market for Shannon theory, 1960–73. J Manage Stud 45(8):1506–1532
Williams LS (2008) The mission statement. A corporate reporting tool with a past, present and future. J Bus Commun 45(2):94–119
Wright M, Clarysse B, Lockett A, Knockaert M (2008) Mid-range universities' linkages with industry: knowledge types and the role of intermediaries. Res Policy 37(8):1205–1223
Wright M, Pruthi S, Lockett A (2006) International venture capital research: from cross-country comparisons to crossing borders. Nottingham University Business School, Nottingham
Zucker LG (1987) Institutional patterns and organizations: culture and environment. Ballinger, Cambridge

IT Support for University Spin-Off Companies

Dragan Mišić, Miodrag Manić, Miroslav Trajanović and Nikola Vitković

Abstract Using information technologies can contribute significantly to the successful start of a company and its later progress. However, successful implementation is connected to certain expenses, which smaller companies (such as university spin-off companies) usually cannot afford. The analysis of existing approaches to implementation of IT and information systems will be performed in this chapter, along with the study of applicability of their use in university spin-off companies. As a conclusion, we will propose the best way to choose and use IT systems.

Keywords University spin-off · Information system · Information technologies

1 Introduction

The modern economy is globalized and knowledge based. Under such conditions, the long-term development of certain regions and countries depends on the competitiveness of their companies. The competitiveness depends on various factors, but one of the most important is the innovation.

The application of innovation in industry is the result of numerous factors, but one of the most significant is government initiation, as well as the orientation and engagement of educational institutions. Among educational institutions, universities play the most important role because their mission is the creation of knowledge. In addition, via their employees, they possess unique knowledge that can and should be used for the development of the economy.

Until the mid-20th century, the traditional mission of the universities was emphasized. It included teaching and the generation of knowledge. In recent decades, however, the contribution of these institutions to economic development is increasingly being mentioned as another one of their missions. In accordance with

D. Mišić (✉) · M. Manić · M. Trajanović · N. Vitković
Faculty of Mechanical Engineering, University of Niš, A. Medvedeva 14, Niš 18000, Serbia
e-mail: misicdr@gmail.com

J. Peña Dopazo and F. Zivic (eds.), *Supporting University Ventures in Nanotechnology, Biomaterials and Magnetic Sensing Applications*,
DOI 10.1007/978-3-319-61237-9_5

these ideas, the universities are beginning to insist on the commercialization of the knowledge they possess. This process initially began in the USA, but it is gradually moving to Europe and other parts of the world.

The commercialization of the knowledge which the university has may be conducted in various ways. These ways include joint research projects with industry, the development of incubators and science parks, patenting and licensing and also the generation of spin-off companies.

There are many different definitions of university spin-off companies in the literature (Pirnay et al. 2003; Druilhe and Garnsey 2004; O'Shea et al. 2008). Those definitions are important because they determine which companies will be taken into account when various analyses and studies are being performed. The most prominent definition is new companies that are based on the technology developed at the university, whereby that technology does not have to, but it can, be protected by patent or license, while the technology developer may, but does not have to, be the entrepreneur which operates the company (Goldstein et al. 2013).

There is a large number of studies related to the success of university spin-off companies. The studies are mainly focused on the economic factors that affect the way the company does business and the survival of these companies, to a greater or lesser extent. One of the factors, which might not be crucial when it comes to the survival and growth of these companies, but might prove helpful, is the use of information technologies. Research in the area of the influence of the use of IT on the functioning of university spin-off companies is rare. On the other hand, there are many studies that refer to the application of IT in small and medium companies. The academic spin-off companies are usually small companies, and, therefore, the conclusions of those studies may also be applied to them.

The idea behind this paper is to scrutinize the need of information processing by university spin-off companies that can be solved through various IT tools. It is performed on insights into the most frequent obstacles that arise in these companies. Finally, we attempted to make recommendations to overcome the obstacles (or at least make them more bearable).

The paper is structured as follows: in Sect. 2, we discuss most common issues related to companies' growth and activities. In Sect. 3, we propose alternative solutions for each of the areas where difficulties were identified. Those solutions are related to the specific software applications.

In the last section, we will make the concluding remarks.

2 Obstacles in the Work of the Spin-Off Companies

The analysis of the business and the growth of academic spin-off companies may be performed from various standpoints. Nowadays, the most mentioned ones are resource-based views and knowledge-based views.

It may be said for a company that it takes resources as an input, and then converts them into products and services that may generate revenue (Barney 1991, 2007).

From the standpoint of resources, a company might be defined as the collection of resources and capabilities that enable it to gain an advantage over the competition. This kind of spin-off company lacks resources by its nature, a lack that consists mainly in non-technical knowledge and skills and starting capital.

On the other hand, the knowledge-based view posits knowledge as the crucial resource of a company. This means that the company ought to focus on the development, protection and transfer of knowledge (Henderson and Cockburn 1994; Liebeskind 1996).

From any point of view, it may be said that the main problems with the development and growth of academic spin-off companies lie in their lack of quality resources and knowledge which would, if present, enable efficient company functioning.

In Geenhuizen and Soetanto (2009), the authors analyze the factors that may lead to lesser growth of a company. Those factors are called obstacles. The authors concluded that in the various stages of development, different types of obstacles occur.

In the first years, the companies encounter problems with cash flow, marketing and management. Furthermore, accommodation and infrastructure might not be resolved properly, although this is less common. In the stages of growth that follow, the number of such issues decreases.

The authors divided the detected difficulties into ones related to marketing, management, finance, physical ones and ones related to the government. Those related to marketing refer to the lack of knowledge and experience in marketing, i.e. the lack of selling skills and an insufficient customer base. The obstacles related to management refer to dealing with uncertainty and overload of the managers. Those related to finance refer to cash flow, investment capital, the lack of research capital etc. Accommodation, infrastructure, distance etc. can be classified as physical issues. The problems with the government refer to various regulatory and the bureaucratic issues.

Wang (2016) examines the barriers to the growth of SMEs. The five most significant perceived by SMEs manages were identified as *access to finance, tax rate, competition, electricity* and *political factors*. Among them, *access to finance* appears to be the greatest barrier, followed by *competition*. Therefore, it is possible to conclude that the barriers resemble those also occurring in academic spin-off companies.

On the basis of the detected difficulties, the possibility of using various information systems (IS) may be considered. The IS would act as a tool that would at least reduce, if not eliminate, these difficulties.

Manager overload would be reduced in the case that the process organization of a company is established (as opposed to the functional), and the business-process management systems are used as to help management.

The issues related to marketing could be solved if the visibility of a company was significantly improved by the establishment of a company web site, after which some of marketing information systems may be used as a marketing aid.

As a support for establishing closer relationship with customers, it is possible to use Customer Relationship Management systems.

In addition to all these types of information systems, comprehensive systems, such as ERP (Enterprise Resource Planning) systems, may also be beneficial. However, it is unlikely that these systems will be implemented in this kind of company, primarily due to their cost and complications linked to their use.

The aforementioned information systems may assist companies in solving problems. Due to the fact that academic spin-off companies are usually small and micro companies, it is unrealistic to expect that large amounts of money can be provided for the purchase of these systems. One feasible solution is to use less expensive solutions, such as those which are based on the open-source philosophy. The services that are offered on the basis of the paradigm Software as a Service (SaaS) are also affordable. This paradigm includes rental of software, which does require payment. The software is placed on the servers of specialized companies, and therefore it is not located within the company itself.

In accordance with all that has been mentioned, in the following section, we provide an overview of existing software in the listed domains. In addition, we make starting recommendations for the use of this software in academic spin-off companies.

3 The Use of IT and Various Types of Information Systems in University Spin-Off Companies

The use of IT in university spin-off companies is a topic which, according to our findings, researchers have not discussed enough. Due to the lack of literature in this domain, we turned to the studies of the use of IT in small and medium companies. This kind of spin-off company can normally be classified as small and or medium companies. There are certain studies that show that the number of employees in these companies, especially in countries other than the USA, remains small (less than 10 employees) even after a few years. Therefore, university spin-off companies are classified as SME (Small and Medium Enterprises).

These companies have peculiarities that distinguish them from other spin-off companies. Those peculiarities are mainly reflected in the fact that the founders of such companies are entrepreneurs who are connected to the faculties and that the companies usually use highly innovative technologies. In our opinion, all of these differences between university spin-off companies and other SMEs do not affect the need for IT in companies. Therefore, the considerations related to SMEs are also related to academic spin-off companies.

The use of IT is one of the important factors that may enhance the efficiency and the competition of enterprises (Henderson and Venkatraman 1993; Kalkan et al. 2011; Peppard 2010). On the other hand, studies show that investment in the newest IT cannot bring benefits by itself (Hussin and Suhaimi 2011). Studies show that the advantage of the use of IT may be gained only in cases where the IT strategy corresponds to the overall business strategy of a company (Almajali and Dahalin 2011; Luftman et al. 1999). In addition, it is necessary for the end users (company employees) to possess the skills required for the use of such systems (Ihlstrom and Nilsson 2003).

When it comes to the application of IT, there is a great difference between small and large enterprises. In comparison with large companies, SMEs have to deal with the limitations in terms of resources and time that can be devoted to the use of IT. Their cost is certainly the greatest barrier, but it is not the only one. The time required for the training and learning of key employees may affect SMEs (and university spin-off companies), because the employees cannot devote themselves to their regular work during that period (Ihlstrom and Nilsson 2003). It may be said that the introduction of IT in SMEs is greatly affected by factors such as anticipated savings, the pressure that arises due to the competition, suppliers and customers, organizational preparedness and ease of use (Grandon and Pearson 2004; Mehrtens et al. 2001). In small companies, such as academic spin-off companies, a small number of people make decisions, the standard procedures have not been instituted yet and long-term planning is limited. Despite all this, the investment in IT may enable long-term survival of the company. Information technologies provide access to external knowledge and financial resources and generate more social-network ties (Morse et al. 2007).

According to Pomffyová and Bartkova (2016), the level of information usage affects business growth, particularly in the sector of micro enterprises. The most common case is the negative impact due to incorrect structure of available information or insufficient analysis. On the other hand, information may ensure a competitive advantage. The most effective tools for achieving this advantage are marketing information systems and business intelligence systems.

3.1 Marketing Information Systems

We have already mentioned that part of the problems that occur in university spin-off companies happens due to the manager's lack of skills in the marketing domain. Another factor that could explain the lower growth of this kind of companies is the great attention that managers pay to technological aspects and their belief that the technology is easy to promote and commercialize. In this manner, the role of marketing is reduced to the level of a tactical activity. Due to the fact that academic spin-off companies base their work on university studies, these companies believe that the superiority of technological innovation is sufficient to affect customers' wishes, thus influencing them to choose a certain product. This leads to a simplified application of marketing principles, often as a short-term response to emerging issues, instead of considering marketing to be a rigorous process of planning. The complexity and turbulence of the high-tech environment (such as university spin-off companies) lead to the conclusion that technological superiority is not sufficient for the success of high-tech products. This means that these enterprises also have to integrate their technological competencies with the marketing capabilities if they wish to understand the customers' needs, assess competitors' strategies and competitive environments and achieve marketplace success (Dutta et al. 1999; Baker and Sinkula 2005).

Market orientation may enhance business performance, especially in high-tech companies that deal with constant and rapid changes in the environment. These companies thus need to be able to use their resources differently in order to satisfy the customers' needs and beat the competition. Although the market orientation ought to positively affect the business of any enterprise, in university spin-off companies, it is even more important because their capability to implement a market orientation is less diffused. As such, the ability to adopt and implement a market orientation should confer on university spin-off companies a sustainable competitive advantage (Abbate and Cesaroni 2014).

Information technologies cannot either improve marketing skills independently, or make the manager start thinking in a manner that would improve the marketing, but they can assist in enhancing the visibility of a company, carrying out its internationalization and performing the marketing function in a consistent manner.

At the beginning of development, the company needs financial assets, market recognition and resources. The capabilities of a company should develop as quickly as possible, and the development should not require a large amount of money. If the enterprise succeeds in getting support from the market and the financial community, its chances of surviving the starting period improve.

There are various information technologies that may be used in the starting phase. The most affordable solution is to establish a web site. The web site brings many benefits to a company, such as reducing the costs of transactions, enhancing advertising and distribution, eliminating the intermediaries, shortening the duration of transaction time and the possibility to monitor the customers' choices, thus achieving market intelligence. Other benefits include the ability to deliver timely information to stakeholders and to offer customized advertising, promotion, and customer services. Finally, web sites provide certain operational benefits related to reduced clerical errors, overhead costs and faster response to new market opportunities.

As we mentioned before, the establishment of a web site is an affordable solution. The cost of creating and maintaining a regular static site is not higher than a few hundred euros. In the case of a dynamic site, the cost is a little higher, and depends on what the company wishes to place on a web site. However, that cost is still relatively low.

In subsequent stages of development, the company's needs for information technologies change. As we have mentioned previously, one of the problems that occur in these spin-off companies is the manager's lack of skills related to marketing. These skills cannot be obtained only by attending courses. Instead, more specific training and gathering of experience are required. The information technologies in this domain cannot teach the manager how to do the marketing- related job, but they can assist her/him to do that job faster and better. For this purpose, it is possible to use so-called Marketing Information Systems.

Marketing Information Systems are intended to bring together disparate items of data into a coherent body of information. Such a system is more than raw data or information suitable for the purposes of decision making. A Marketing Information System has traditionally been proposed to provide marketing managers a thorough process of intelligence gathering. These systems also provide methods for interpreting the information they provide. According to Kotler (2009), the definition of such systems may be "a marketing information system consists of people, equipment and procedures to gather, sort, analyze, evaluate and distribute needed, timely and accurate information to marketing decision makers".

Marketing information systems consist of an internal reporting system, marketing research system, marketing intelligence system and various marketing models.

Companies that have worked for a long time contain a great deal of information. This information is not always used properly, because it is often divided between various departments and/or managers. It often happens that the staff that works in one department and possesses certain information cannot imagine how that information would be helpful to the decision makers in other areas. Even the decision makers themselves are not able to predict which information from other areas is required for proper decision making, so they do not request it.

Internal data that can help the decision makers are, for example, received orders, exiting stocks and sales invoices. Those are only a few examples of internal data that the marketing manager may use, but even they can be used for obtaining a lot of new information. For instance, it is possible to obtain information about types of products according to territory, industry, customers, average values and net sales according to the same parameters.

Marketing research refers to a proactive search for information. The company that orders such studies is attempting to solve a perceived problem. In most cases, the data collection is directly dependent on the domain. There are other forms of marketing research that are, for example, focused on continual monitoring of the marketing environment.

A market intelligence system is a set of procedures and data sources that marketing managers use to extract data from the environment. These data are later used for decision making.

Marketing models represent means by which the information is interpreted, with the aim of making a decision. These models may or may not be digitalized. Marketing models are analytical subsystems of the whole marketing information system.

Marketing automation software and Customer Relationship Management (CRM) systems are often considered to be the same type of application. In general, it may be said that the main difference between them is who they target. CRM is focused on sales, while the Marketing automation software is focused on marketing. The common feature of these systems is that they use the same information (email, address, phone number, company name etc.).

Marketing automation enables firms to follow a prospect's top-of-funnel activities, such as when they visit company's website, open an email, read a company's blog or fill out a form. Marketers commonly use marketing automation software to schedule and track marketing campaigns, especially email campaigns and mass business-to-customer communications. Some of the benefits of marketing automation software are (Justen 2014):

- The ability to easily segment prospects into the appropriate mailing lists based on their past interactions with the company or their interests and preferences.
- Lead nurturing functionality, which enables firms to automatically send triggered emails at the time when a person is most interested in a company's products or services. A company can also schedule a series of emails in a "drip" campaign, so that the company stays top-of-mind with its prospects.
- Once a campaign has ended, the system generates analytics showing how successful the campaign was.

How does one pick software for marketing automation? There are specialized software applications on the market that are used exclusively for marketing automation. Marketing automation software may be divided into commercial and open-source applications. In general, the commercial applications usually offer more features than the open-source applications, but the latter have been improving and getting closer to the commercial ones.

It is likely that small and micro enterprises, such as academic spin-off companies, will not be able to spend enough money on the purchase of commercial products related to marketing automation. It is far more probable that they will turn to open-source solutions. Open-source solutions are not free, but their cost is much lower than that of the commercial programs.

In the following part of this section, we mention some of the commercial solutions that may be used for marketing automation. This list is neither exhaustive nor exclusive, and it was composed on the basis of authors' current insights into marketing automation software. One should think of it as an illustration and the starting point for the research that a company ought to conduct in the case that is planning to purchase this kind of software.

When it comes to commercial software in this area, the companies and products that are often mentioned are Infusionsoft, Hubspot, Ontraport, Marketo, Pardot, Eloqua etc. The choice of solution depends on the requirements that the software needs to satisfy and its simplicity of use. There are many factors that should be taken into account, and there are many web sites on the internet that can provide advice when choosing software for marketing automation. The cost of this software ranges from a few hundred to a few thousand dollars a month, depending on the company's needs and the number of customers.

Some of the aforementioned software solutions are put to best use in large companies, while some are more scalable, so they can also be used in smaller companies. For instance, a list of marketing automation software whose starting price is very low or even zero free is shown on the web site (Mosley 2016). The price of these solutions increases along with the needs, i.e. the number of contacts and similar characteristics. Leadsius, Sales Autopilot, Azuqua, Spokal, JumpLead, Nurture, Infusionsoft and Genoo are cited on this list. Some of these applications are completely free at first, so they are the most suitable solution for small and micro enterprises, such as academic spin-off companies.

3.2 Customer Relationship Management Systems

In the first stages of growth of academic spin-off companies, it is not easy to attract customers to purchase their products. Those products are usually highly innovative, so there is no customer type that regularly purchase them. It may be said that the enterprise defines a new type of customers at that moment.

In the later stages, however, they face the same problems as other smaller companies. Thus, in Geenhuizen and Soetanto (2009), the authors state that the fall in the reduction rate at age four reflects the failure to gain stable relations with customers and the liability in becoming independent in gaining key resources. One approach that might prevent the decline in reduction rate is the use of automated tools (such as CRM) to maintain the relationship with the customer.

It is difficult to define Customer Relationship Management systems. There are many various definitions that arise from emphasizing the various aspects which CRM deals with. One of the most comprehensive ones is that "CRM is the core business strategy that integrates internal processes and functions, and external networks, to create and deliver value to targeted customers at a profit. It is grounded on high quality customer related data and enabled by information technology" (Buttle 2008).

The enterprise uses CRM data for optimization of customer interactions in order to increase sales and customers' satisfaction. The introduction of a CRM system may bring many benefits. The customer is nowadays very valuable to every organization, because there is immense competition in the market. The company may use CRM to satisfy their requests in order to establish long-term relationship with them. It may be said that the primary condition for further growth of a company is the customers' satisfaction.

Implementation of CRM enables information about customers to be shared within an organization (Chen and Popovich 2003; Zablah et al. 2004). It also enables a company to define various types of clients that it will serve in various ways. When there is sufficient information, CRM enables an organization to focus on the most profitable customers.

For example, CRM enables the staff that specializes in sales to see where in the process of sales the customer is at the moment and to help him/her close the deal. CRM also enables sending of promotional material, establishing a clear idea of who the individual that the staff is communicating with is, whereby the conversation becomes more personal and successful. It enables sending internal alarms when something client-related happens (his/her account is changing, it is her/his birthday, etc.) Simply put, CRM systems help to secure sales by making the sales process a more personal experience for the customer.

Despite all the mentioned benefits that CRM could bring to a company, many small companies hesitate to introduce these systems. There are numerous reasons for this. For example, Ramaseshan and Chia (2008) found that small enterprises do not introduce CRM due to the cost of hardware and software. It is common practice to evaluate what could be gained by implementing the new technology and to compare the possible gains to the cost of introduction. When introducing the CRM, it should be taken into account that the cost of maintaining the customer base might be high. A report of the Gartner group stated that about 55% of all CRM projects fail, and these failures 20% affect long-standing relationships (Mello 2002). This is because they have less understanding about CRM technologies and do not know how a particular CRM approach should be cost-effectively implemented (Loh et al. 2011).

In addition to not knowing enough about the basic concepts, which leads to termination of the use of CRM, managers are also often not able to estimate where to start and which technology and system to use. Before introducing the CRM, the managers should know precisely which type of CRM is suitable for their enterprise and thus reduce the possibility of unsuccessful implementation.

In this section, we mention some of the potential software solutions related to CRM, as we did in the previous one. We now discuss the so-called open-source solutions, primarily because of the overall lower price.

Although it is thought that open-source is the same as cost-free, that is not always the case. Open-source tools may or may not be free, but it is still true that open-source solutions cost less than commercial solutions. This is particularly important in academic spin-off companies, which usually do not possess sufficient funds for the purchase of corresponding software solutions. Although the price that is to be paid is relatively low, other issues may occur, such as with availability of human resources. These spin-off companies may face this kind of problems, because of their lack of human resources at the beginning of development. The possible solution is to leave the introduction and maintenance to third-party companies, but, by doing this, the initially-low cost increases.

Another advantage of open-source solutions derives from the fact that the company receives the source code of the software they purchase. This means that various adaptations are possible. Commercial software does not usually provide this possibility.

When discussing free software, it should be mentioned that there are often various limitations in the free versions of software. In CRM software, those limitations refer to the number of users, the number of contacts and the size of storage. The number of users ranges from two to five, which might be perfectly suitable for small enterprises such as academic spin-off companies. When it comes to the number of contacts, there is no uniformity because some products count contacts, while others count all records related to the contacts.

There are many articles related to open-source CRM tools on the internet. They represent a good starting point for making decisions about which software to purchase. When deciding, it is not only the software characteristics that should be taken into account, but it also needs to be thought about what the basic version offers and the type of support that the company offers. There are solutions that have poor support, and it thus becomes challenging to install, configure and use the software. For example, Wood (2014) provides a list of solutions that she considers to be the best in the domain of CRM solutions with low or without any initial outlay. The author mentions open-source as well as CRM solutions whose starting version is free. Those are SuiteCRM, vTiger, Zurmo, CapsuleCRM, Insightly, Really Simple Systems, FreeCRM, Bitrix24, Raynet and ZohoCRM. These are just a few possibilities, but they represent a starting point for studying this type of software.

We mentioned earlier that Marketing Information Systems and CRM systems use similar or the same data, which they then interpret in different ways. This suggests that it is best to look for solutions that enable integration of these systems, or in which these components are already integrated.

Recently, the application of social networks for business purposes has become more frequent. Social networks are no longer just a way of communication among individuals, but a communication channel in enterprises that leads to business growth (Rahman et al. 2016). This particularly applies to customer support, so that social networks are today an integral part of contemporary CRM systems.

3.3 Business Process Management Systems (BPMS)

One of the problems that academic spin-off companies face is management. Most papers that discuss this topic emphasize the insufficient managerial skill of the owner and leaders of the enterprise (Geenhuizen and Soetanto 2009; Vohora et al. 2004). In addition to these skills, people in these enterprises often lack the time for operating the company management. Business Process Management Systems may be very helpful when dealing with the latter.

Business Process Management (BPM) is a structured method of understanding, documenting, modeling, analyzing, simulating, executing and continuously changing end-to-end business processes and all relevant resources in relation to an organization's ability to add value to the business.

The most important insight that characterizes BPM is a 'process view' of management that eschews the functional boundaries of an organization's various departments in favour of a more 'holistic' approach (Baker and Maddux 2005; Rosemann and de Bruin 2004). Each step of the production value chain, from supplier to customer, can be monitored and explicitly linked to corporate strategy, operational efficiency and competitive advantage (Harmon 2003).

The benefits of a successful process-improvement effort include: better operational efficiency; increased profitability; better customer relations; shorter process-cycle times; lower operating costs; increased accountability; and improved market competitiveness (Ahadi 2004; Raymond et al. 1998).

There are various views on BPM coming from various different domains. Some think of it as a management strategy, while IT specialists consider it to be a software system. There is still no common consensus about the definition of Business Process Management itself (Aalst et al. 2003).

In this review, we discuss BPM in the context of the information systems that are used as its support. BPMS may be defined as a "(suite of) software application(s) that enable the modeling, execution, technical and operational monitoring, and user representation of business processes and rules, based on integration of both existing and new information systems functionality that is orchestrated and integrated via services" (Ravesteyn and Versendaal 2007).

Are BPM tools essential to small enterprises, such as academic spin-off companies? The answer is that they are probably not as necessary in small companies as they are in larger ones. When it occurs in a small company that it is required to adjust the process or solve arising problems, solutions are found on the run, and company operations continue. This is quite easy to do in the case that the number of processes is small and the issues rarely occur. However, this way of doing business limits the growth of a company. When a business starts to grow, and the processes by which it functions are ineffective (or not even defined), it is likely that the problems are going to occur.

Various studies that refer to the application of BPM in small companies show that these companies may increase profits by using this concept and appropriate tools. For example, in Dallas and Wynn (2014), the authors conclude that process models and the documentation that goes with them may improve the efficiency of client-service implementation and ongoing delivery. The companies have also managed to gain a consistent, shared understanding of how work is undertaken and have provided a means for capturing the procedural information that would otherwise be retained only by individuals as tacit knowledge. SMEs usually do not need complex solutions with full-blown automated BPMS, but, as the first step, well-defined measurable processes eventually supported by lightweight tools are enough.

However, it is not easy to introduce a BPM. The studies that refer to the introduction of BPM in small enterprises show that there are many problems that must be solved and mistakes that must be avoided when introducing it. For example, Chong (2007) concludes that the implementation of a BPM in small enterprises is not straightforward, due to the lack of finance and time. This conclusion applies especially to academic spin-off companies that have below 20 employees, which is also the case with the companies highlighted in the paper.

In Imanipour et al. (2012), the authors divide the factors that influence inhibiting the implementation and adoption of BPM in SMEs into organizational, environmental, technological and individual. Some of the organizational factors are: lack of clarity on a strategic level; lack of time and the pressure stemming from it; lack of human resources and multiple roles of employees; lack of financial resources and the pressure stemming from it; poor knowledge of process oriented approaches; lack of BPM education in employees; etc. The environmental factors include the difficulty of finding and retaining qualified personnel with required skills and knowledge, lack of companies offering BPM software and services, competitive pressure etc. The technological factors are: lack of defined business metrics and/or measurement protocols for assessing process management performance; lack of adequate Information Technology infrastructure; perceived gaps between processes design (BPM implementation) and process execution (BPM execution); lack of simple procedures and guidelines about how to start the process; etc. The individual factors are: lack of Information Technology Expertise (in top management); lack of a standard language and a common understanding on quiddity of process orientation and BPM (concepts, technologies and tools) between experts, managers, companies and employees; the absence of a cross-functional mindset amongst top management; lack of BPM project support in implementation and adoption (supervisors, top management); etc.

Based on the aforementioned, it may be concluded that small enterprises (and academic spin-off companies) should consider implementation of BPM systems. It should be noted that the costs of these tools is several times higher than the cost of tools that are needed for marketing automation or a CRM solution. Therefore, we recommend the use of less expensive solutions, based on open-source tools. When deciding, it should be taken into account that not only is cost important, but also the offered characteristics. It should be decided earlier which characteristics a specific BPMS implementation should include.

When it comes to applications of BPM systems, it is important to emphasize that their limiting factor might be support for processes that are changed often. Until now, BPMS proved valuable for strictly defined processes. In the general case, it is not always possible to represent a certain process in such a manner that it does not need to be altered. This is especially true for young companies which are still in the initial phase of defining processes. Their processes need to be constantly adapted and adjusted. That means that BPMS should enable flexibility. Various systems have different approaches in order to solve this problem. Some of those approaches refer to applications of artificial-intelligence methods for process adjustments. For example, for that purpose, Misic et al. (2010a) used an expert system connected to BPMS to detect exceptions and adjust processes. In Misic et al. (2010b), the ASM developed at the Faculty of Mechanical Engineering in Nis is used for exception recognition. ASM is a type of artificial-intelligence system that is able to autonomously make conclusions using sematic relations between various concepts. ASM is described in more detail in Stojkovic et al (2015), Trifunovic et al. (2015a, b). There are other systems which could provide the flexibility for BPMS. For instance, Adams et al. (2007) used the worklet technology; in Reichert and Dadam (2009), the authors introduced a new language which enables description of dynamical blocks that can used to build a process; other authors use ontologies, e.g. Yao and Kumar (2013) and Dang et al. (2008).

University spin-off companies are most often enterprises where knowledge is of the greatest importance. Their processes are therefore knowledge intensive. For convenient management of such processes, a new type of systems is used, namely Adaptive Case Management Systems. The central component of these systems is a knowledge worker who should be able to execute processes in accordance with the current situation and not in accordance with a single predefined process. In those ad hoc processes, a person decides the next steps, and the system should assist in that. Those systems are depicted in more detail in Motahari and Swenson (2013).

We will mention again that there many papers on the internet that compare the commercial BPM tools and the open-source ones; also, there are studies which were presented at various conferences. Based on all this, we recommend tools such as YAWL and JBpm, which may be installed for free. There are more tools that developed as open-source, such as Bonita, ProcessMaker etc., but, in comparison with previous two, those have higher initial prices.

A broad overview of open-source BPM tools can be found in Meidan et al. (2017). This paper can provide a starting point for the enterprises intending to utilize less expensive, open-source tools. The most advantageous tools mentioned there are YAWL and jBPM. Other tools included in this review are Bonita, ProcessMaker, Camunda, Activity and uEngine.

3.4 *Cloud Computing*

From what can be seen in previous literature related to the use of specific information systems in academic spin-off companies, it may be concluded that one of the basic problems with the application and implementation of these systems is the lack of resources, both financial and human. In addition to the already recommended ones, one of the possible solutions is cloud computing.

There are many definitions of cloud computing; one that is comprehensive is from Vaquero et al. (2009). According to the authors: "Clouds are a large pool of easily usable and accessible virtualized resources (such as hardware, development platforms and/or services). These resources can be dynamically reconfigured to adjust to a variable load (scale), allowing also for an optimum resource utilization. This pool of resources is typically exploited by a pay-per-use model in which guarantees are offered by the Infrastructure Provider by means of customized Service-Level Agreements".

Cloud-services delivery models can be broadly categorized into four types: Software-as-a-Service (SaaS), Service, Platform-as-a-Service (PaaS) and Infrastructure-as-a-Service (IaaS) (Lin and Chen 2012). We are interested in the first two models, SaaS and Services, which may be useful to us. These models are based on the fact that the customers should pay, rent or subscribe to applications or services from the cloud providers to access applications or services via the internet (Leavitt 2009).

The importance of cloud computing has been acknowledged by the European Union which finances projects that apply cloud computing in SMEs. In Assante et al. (2016), one of such projects is depicted: its most useful characteristics according to the authors are on-demand self-service, broad network access, resource pooling, rapid elasticity and measured service.

How is cloud computing (SaaS) connected with the information systems that are recommended for use in academic spin-off companies? All proposed solutions (Marketing Information Systems, Customer Relationship Management Systems, and Business Process Management Systems) are either already available in this form or they will be, in the near future. These solutions must be considered by this kind of company, as well as other small enterprises.

Before opting for this kind of solution, the managers ought to consider the advantages and disadvantages of SaaS model, especially from the aspect of small enterprises. There are many empiric studies that attempt to look at the use of cloud computing from the perspective of small and medium companies.

The benefits that cloud computing bring to small and large enterprises are similar, but those benefits are more evident in small companies. This primarily refers to lower price, since the initial investments in hardware and software are much smaller, as well as the maintenance costs. Also, the costs are usually paid by model pay-per-use (Vaquero et al. 2009), which means there are no big investments all at once, instead, the payments are rather small and occasional. Besides, this way of paying enables small enterprises to access software that would otherwise be too expensive for them. In addition, in Haselmann and Vossen (2011), the authors conclude that the advantages that SaaS model offers are handling data/storage related tasks by the service provider, better scalability and software performance, delegating IT infrastructure maintenance to the service provider, no license or software issues when replacing hardware, no software maintenance etc. Also, SaaS is ready for use more quickly than traditional software.

In Haselmann and Vossen (2011), the authors note potential disadvantages of introducing SaaS model in small companies. Those are primarily integration of SaaS with existing company software, loss of control over access to data, too-tedious data migration from existing application, high latency when using software (page loads), less control over backups of data, SaaS software lacks some desired features, permanent broadband internet connection required etc.

Factors that may influence the decision to use cloud computing in SMEs (Ross and Blumenstein 2015) are:

- The lack of understanding of the infrastructure, cost and appropriateness to the needs and scenarios of various companies from different business environments.
- The ICT-skill levels of users, managers and entrepreneurs.
- The readiness of SMEs to adopt cloud computing from a business perspective.
- Less time.

The aforementioned advantages and disadvantages indicate that a thorough analysis should be performed before using this solution. However, cloud computing has a good perspective, and it is one solution that must be considered.

4 Conclusion

Academic spin-off companies are increasingly gaining in importance. The universities are gradually changing their goals and, in addition to educational institutions, they are becoming a place where new economic entities come to life. In the world of science, there are many studies that analyze these companies from various aspects. They are mostly researches that attempt to improve models of development of spin-off companies or to explain the reasons for their success or failure.

The studies that refer to the use of information technologies in academic spin-off companies are almost nonexistent. On the other hand, there are many studies related to the application of IT and various types of information systems in small and medium companies. The majority of spin-off companies can be categorized as small, even micro, enterprises. Therefore, the conclusions drawn about SMEs may be applied to university spin-off companies, regarding certain specificities of the latter.

We made a connection between the studies related to SMEs and university spin-off companies on the basis of obstacles that were spotted in the latter. These obstacles affect the companies' growth and prevent their further development to a lesser or greater extent. In addition to proposed solutions of existing studies, we believe that the introduction of various types of information systems may assist in their reduction or elimination.

Since academic spin-off companies are often linked with marketing, and the establishment of stable relationship with customers and management, we discussed the possibility of the application of information systems from appropriate domains. We thus recommended the use of Marketing Information Systems, Customer Relationship Management Systems and Business Process Management Systems.

In accordance with the limitations that exist in academic spin-off companies, and that refer to the lack of financial and human resources, we proposed solutions that will minimize these needs. It is clear that it is impossible to eliminate all expenses, but they might be reduced to a reasonable level. Therefore, the introduction of an information system would return more to a company than it is needed to invest in implementation and maintenance of the system.

In addition to the types of information systems we discussed, there are other types of specific or comprehensive information systems that could be helpful to companies. This primarily refers to Enterprise Resource Planning (ERP) systems, as one of comprehensive solutions. We did not discuss these systems for several reasons. Studies show that academic spin-off companies, especially in Europe, usually employ less than 10 people, which means they belong among the micro enterprises. Such companies generally do not require ERP systems, and it is unlikely that they have enough resources for purchasing these systems. ERP systems are usually very expensive solutions, with a specific interface. Therefore, in addition to the financial resources required for their purchase, it is desirable for companies to have people at their disposal who would deal only with them.

The review of literature related to this domain shows that the application and influence of IT and information systems on the development of university spin-off companies have not been well considered in scientific community. This review may be an invitation for scientists to study the functioning of these companies, from this perspective.

Acknowledgements The research study is part of project III41017 Virtual Human Osteoarticular System and its application in preclinical and clinical practice, funded by the Ministry of Education and Science of Republic of Serbia, for the period of 2011–2017.

References

Aalst W, Hofstede AHM, Weske M (2003) Business process management: a survey. In: Proceedings of the 2003 international conference on business process management, Eindhoven, The Netherlands

Abbate T, Cesaroni F (2014) Market orientation and academic spin-off firms. In: Indem Carlos III University, working paper business economic Series. ISSN: 1989-8843, 01/2014

Adams M, Hofstede AHM, Aalst W (2007) Dynamic and extensible exception handling for workflows: a service-oriented implementation. Technical report, BPM Center Report BPM 07-03

Ahadi HR (2004) An examination of the role of organisational enablers in business process reengineering and the impact of information technology. Inf Resour Manag J 17(4):1–19

Almajali D, Dahalin Z (2011) Applying the triangulation approach in IT—business strategic alignment and sustainable competitive advantage. IBIMA Bus Rev J 2011:1–13

Assante D, Castro M, Hamburg I, Martin S (2016) The use of cloud computing in SMEs. Second international workshop on mobile cloud computing systems, management, and security, Madrid, Spain

Baker G, Maddux H (2005) Enhancing organizational performance: facilitating the critical transition to a process view of management. SAM Adv Manag J 7(4):40–60

Baker WE, Sinkula JM (2005) Market orientation and the new product paradox. J Acad Mark Sci 33:461–475

Barney J (1991) Firm resources and sustained competitive advantage. J Manag 17(1):99–120

Barney J (2007) Gaining and sustaining competitive advantage, 3rd edn. Prentice Hall, Englewood Cliffs

Buttle F (2008) Customer relationship management: concepts and technologies, 2nd edn. Taylor & Francis Ltd, Oxford

Chen J, Popovich K (2003) Understanding customer relationship management (CRM) people, process and technology. Bus Process Manag J 9(5):672–688

Chong S (2007) Business process management for SMEs: an exploratory study of implementation factors for the Australian wine industry. J Inf Syst Small Bus 1(1–2):41–58

Dallas I, Wynn MT (2014) business process management in small business: a case study, information systems for small and medium-sized enterprises. In: Devos J, Landeghem H, Deschoolmeester D (eds) Information systems for small and medium-sized enterprises. Springer, Berlin, pp 25–46

Dang J, Hedayati A, Hampel K, Toklu C (2008) An ontological knowledge framework for adaptive medical workflow. J Biomed Inform 41:829–836

Druilhe C, Garnsey E (2004) Do academic spin-outs differ and does it matter? J Technol Transf 29 (3–4):269–285

Dutta S, Narasimhan O, Rajiv S (1999) Success in high-technology markets: is marketing capability critical? Mark Sci 8(4):547–568

Geenhuizen MS, Soetanto DP (2009) Academic spin-offs at different ages: a case study in search of key obstacles to growth. Technovation 29(10):671–681

Goldstein H, Peer V, Sedlacek S (2013) Vienna as a region of knowledge: increasing the generation of University spin-offs. Final report to the Vienna Chamber of Commerce (WKW). http://www.modul.ac.at/departments/public-governance-and-sustainable-development/projects. Accessed Sept 2014

Grandon EE, Pearson JM (2004) Electronic commerce adoption: an empirical study of small and medium us businesses. Inf Manag 42:197–216
Harmon P (2003) Business process change: a manager's guide to improving, redesigning, and automating processes. Morgan Kaufmann, San Francisco
Haselmann T, Vossen G (2011) Software-as-a-service in small and medium enterprises: an empirical attitude assessment. In: Bouguettaya A, Hauswirth M, Liu L (eds) Web information systems engineering—wise 2011, Springer, Berlin, pp 43–56
Henderson J, Venkatraman N (1993) Strategic alignment: leveraging information technology for transforming organizations. IBM Syst J 32(1):4–16
Henderson R, Cockburn I (1994) Measuring competence? Exploring the effects in pharmaceutical research. Strateg Manag J 15:63–84
Hussin H, Suhaimi MA (2011) Information technology and business alignment in Malaysia SMEs. IIUM Press, Malaysia
Ihlstrom C, Nilsson M (2003) E-business adoption by SMEs: prerequisites and attitudes of SMEs in a Swedish network. J Organ Comput Electron Commer 13(3):211–223
Imanipour T, Talebi K, Rezazadeh S (2012) Obstacles in business process management (BPM) implementation and adoption in SMEs. Available at SSRN: http://ssrn.com/abstract=1990609. Accessed Jul 2016
Justen A (2014) What's the difference between CRM and marketing automation software. http://blog.capterra.com/whats-difference-between-crm-and-marketing-automation-software/. Accessed Jan 2016
Kalkan A, Erdil O, Çetinkaya Ö (2011) The relationships between firm size, prospector strategy, architecture of information technology and firm performance. Proced Soc Behav Sci 24: 854–869
Kotler P (2009) Marketing management, a South Asian perspective. Pearson Education, New York
Leavitt N (2009) Is cloud computing really ready for prime time? Computer 42(1):15–20
Liebeskind JP (1996) Knowledge, strategy, and the theory of the firm. Strateg Manag J 17:93–107
Lin A, Chen NC (2012) Cloud computing as an innovation: perception, attitude, and adoption. Int J Inf Manag 32(6):533–540
Loh BK, Koo K, Ho K, Idrus R (2011) A review of customer relationship management system benefits and implementation in small and medium enterprises. In: MCBANTA'11 proceedings of the 12th WSEAS international conference on mathematics and computers in biology, business and acoustics, 2011, pp 247–253
Luftman JN, Papp R, Brier T (1999) Enablers and inhibitors of business-IT alignment. Commun Assoc Inf Syst 1(11):1–33
Mehrtens J, Cragg PB, Mills AM (2001) A model of internet adoption by SMEs. Inf Manag 38:165–176
Meidan A, García-García JA, Escalona MJ, Ramos I (2017) A survey on business processes management suites. Comput Stand Interfaces 51:71–86
Mello A (2002) Six mistakes that will sink your CRM. http://www.zdnet.com/article/six-mistakes-that-will-sink-your-crm. Accessed Oct 2016
Misic D, Domazet D, Trajanovic M, Manic M (2010a) Concept of the exception handling system for manufacturing business processes. Comput Sci Inf Syst 7(3):489–509
Misic D, Stojkovic M, Domazet D, Trajanovic M, Manic M, Trifunovic M (2010b) Exception detection in business process management systems. J Sci Ind Res 69(3):188–193
Morse EA, Fowler SW, Lawrence TB (2007) The impact of virtual embeddedness on new venture survival: overcoming the liabilities of newness. Entrep Theory Pract 31(2):139–159
Mosley M (2016) Marketing automation tools for your business. https://referralrock.com/blog/marketing-automation-tools/. Accessed Jun 2016
Motahari HR, Swenson KD (2013) Adaptive case management: overview and research challenges. In: IEEE international conference on business informatics, Vienna, Austria
O'Shea P, Chugh H, Allen T (2008) Determinants and consequences of university spinoff activity: a conceptual framework. J Technol Transf 33(6):653–666

Peppard J (2010) Unlocking the performance of the Chief information officer. Calif Manag Rev 52:73–99

Pirnay F, Surlemont B, Nlemvo F (2003) Toward a typology of university spin-offs. Small Bus Econ 21(4):355–369

Pomffyová M, Bartkova L, (2016) Take advantage of information systems to increase competitiveness in SMEs. In: 19th international conference enterprise and competitive environment 220, pp 346–354

Rahman NA, Yaacob Z, Radzi R (2016) An overview of technological innovation on SME survival: a conceptual paper. Proced Soc Behav Sci 224:508–515

Ramaseshan B, Chia PK (2008) Factors influencing implementation of CRM technology among small and medium sized enterprises. In: Proceedings of Australian and New Zealand marketing academy conference, Sydney, 2008

Ravesteyn P, Versendaal J (2007) Success factors of business process management systems implementation. In: 18th Australasian conference on information systems, Toowoomba, Australia, pp 396–406

Raymond L, Bergeron F, Rivard S (1998) Determinants of business process reengineering success in small and large enterprises: an empirical study in the Canadian context. J Small Bus Manag 36(1):72–85

Reichert M, Dadam P (2009) Enabling adaptive process-aware information systems with adept2. In: Van Der Aalst WMP (ed) J Cardoso. Handbook of research on business process modeling. Information Science Reference, New York, pp 173–203

Rosemann M, Bruin T (2004) Application of a holistic model for determining BPM maturity, AIM Pre-ICIS workshop on process management and information systems (Actes du 3e colloque Pre-ICIS del'AIM). In: Akoka J, Comyn-Wattiau I, Favier M (eds) 12 December, Washington DC, pp 46–60

Ross PK, Blumenstein M (2015) Cloud computing as a facilitator of SME entrepreneurship. Technol Anal Strateg Manag 27(1):87–101

Stojkovic M, Trifunovic M, Misic D, Manic M (2015) Towards analogy-based reasoning in semantic network. Comput Sci Inf Syst 12(3):979–1008

Trifunovic M, Stojkovic M, Misic D, Trajanovic M, Manic M (2015) Recognizing topological analogy in semantic network. Int J Artif Intell Tools 24(3): 1550006-1–1550006-25

Trifunovic M, Stojkovic M, Trajanovic M, Manic M, Misic D, Vitkovic N (2015b) Analysis of semantic features in free-form objects reconstruction. Artif Intell Eng Des Anal Manuf 30(1):1–20

Vaquero LM, Rodero-Merino L, Caceres J, Lindner M (2009) A break in the clouds: towards a cloud definition. Comput Commun Rev 39(1):50–55

Vohora A, Wright M, Lockett A (2004) Critical junctures in the development of university high-tech spinout companies. Res Policy 33(1):147–175

Wang Y (2016) What are the biggest obstacles to growth of SMEs in developing countries?—an empirical evidence from an enterprise survey. Borsa Istanb Rev 16(3):167–176

Wood C (2014) The 10 best free and open source crm software solutions. http://blog.capterra.com/free-and-open-source-crm/. Accessed Nov 2016

Yao W, Kumar A (2013) CONFLEXFLOW: integrating flexible clinical pathways into clinical decision support systems using context and rules. Decis Support Syst 55(2):499–515

Zablah R, Bellenger DN, Johnston WJ (2004) An evaluation of divergent perspectives on customer relationship management: towards a common understanding of an emerging phenomenon. Ind Mark Manag 33:475–489

Differences Between Adopters and Non-adopters of Innovation: Case Study of New Technologies|Adoption by Small and Medium Enterprises in Serbia

Fatima Zivic, Nenad Grujovic and Jasmina Miljojkovic

Abstract This article explores the factors that govern the process of innovation adoption by companies. We have studied the factors that promote the willingness or resistance of companies to adopt new research results or novelty, in general. As one case study, we will elaborate factors that influence a website adoption by smaller companies that mainly do not have their own ICT departments. The study investigated the attitudes of managers towards Internet-mediated business development and promotion, comprising a case study of small and medium-sized enterprises (SMEs) in Serbia. Special focus was placed on low income micro-SMEs. The results showed that, in the case of larger organisations, the models proposed by Flanagin (Hum Commun Res 26(4):618–646, 2000) and Davis (MIS Q 13(3):319–339, 1989) can be effectively used as guidance for website adoption. However, in the case of very small companies, with limited budgets and locally oriented markets, a few of those factors are more prominent, while some have very little influence on the adoption process. Education background of the owners essentially determines decision making and the adoption process, and competitive pressures have no significant influence. The annual income of the company has a certain influence, but, even with the available budget, micro-SMEs will not consider introduction of a website if they cannot perceive any benefits, often due to a lack of awareness of innovations in ICT and marketing. If the owners can perceive website benefits, the cost of introduction and maintenance has little influence. Specific cultural aspects of the region should be considered as the significant factor influencing innovation introduction.

F. Zivic (✉) · J. Miljojkovic
Faculty of Engineering, University of Kragujevac,
S. Janjic 6, 34000 Kragujevac, Serbia
e-mail: zivic@kg.ac.rs

J. Miljojkovic
e-mail: jasmina.miljojkovic@fink.rs

N. Grujovic
Center for Information Technologies, Faculty of Engineering,
University of Kragujevac, S. Janjic 6, 34000 Kragujevac, Serbia
e-mail: gruja@kg.ac.rs

J. Peña Dopazo and F. Zivic (eds.), *Supporting University Ventures in Nanotechnology, Biomaterials and Magnetic Sensing Applications*,
DOI 10.1007/978-3-319-61237-9_6

Keywords Emerging technologies · Small and medium enterprises (SMEs) · Corporate website adoption · Innovation

1 Introduction

According to the EC Commission Recommendation 2003/361/EC, the category of micro-, small- and medium-sized enterprises (SMEs) is made up of enterprises that employ fewer than 250 persons and have an annual turnover not exceeding 50-million euro, and/or an annual balance sheet total not exceeding 43-million euro. Some SMEs are rather large, meaning they can dedicate more significant expenditures to the online business promotion and marketing, which is not always the case when micro-SMEs are considered. This is especially true in the case of low income countries, such as in the Serbian SME sector, consisting mainly of micro-SMEs with very limited budgets and less than 10 employees, using the EU definition of micro as <10 employees and small as 10–49 employees. According to the official statistics, 98.9% of active enterprises in Serbia have been categorized as SMEs (Statistical Office of the Republic of Serbia 2011). A similar situation is present mainly in the whole Balkan region (Dimic and Pavlovic 2016). In most of the cases, such small firms have very low budgets for promotional activities and very often consider online marketing as a cost-effective form of promotion.

One of the recognised issues in the case of small companies is a lack of resources to follow the latest innovations in various areas, even in those closely related to their business (Rizos et al. 2016; Eito-Brun and Sicilia 2016). Several studies indicate that networking and collaboration with research-performing institutions, such as faculties or research institutes, can immensely improve innovation in companies (Džupka et al. 2016; Estensen and Ekambaram 2016; Lamontagne 2015; Gottfridsson 2014; Tajonar 2014; Kittilaksanawong and Ren 2013; Scarpellini et al. 2012). However, to some degree, companies, especially small ones, show certain resistance towards researchers and innovations, even in the widely adopted information and communications technologies (ICT). ICT solutions have enabled the establishment of the online community which is increasingly becoming a world unto itself. Persons often create a parallel identity not necessarily identical to their actual appearance in the real world. However, a company should create its online identity profoundly based on its real business status. Promotional campaigns on highly rated TV networks or newspapers are very expensive, especially if compared to Internet-mediated promotion. The cost of communication using Internet technologies is significantly lower, and, at the same time, it can reach out to an extensively large number of potential customers, if properly deployed. However, not all the companies have satisfactory online promotion, mainly due to the lack of an adequate approach.

Maybe the most acknowledged way to promote the company online is a corporate website. Attracted by the links, banners, text ads or some other form of online promotion, Internet users will visit the website, but not always search through it or come back to visit it again. Knowledge acquisition is often blocked by poor functionality of the website, and the visitor leaves the site dissatisfied (Fang and Holsapple 2007). Besides functionality, content is a common problem. According to Kang and Kim (2006) quantity will only keep low-interest visitors on the website, while high-interest visitors are more interested in well-organized content. There are many criteria that could be used to evaluate website quality (Lian et al. 2016; Faustina and Balaji 2016; Hasan and Abuelrub 2011; Lee and Kozar 2006; Hernández et al. 2009; Kim and Niehm 2009; Kim and Stoel 2004; De Marsico and Levialdi 2004; Webb and Webb 2004; Lombard and Snyder-Duch 2001).

Essentially, websites can be developed as product-oriented or corporate-oriented (Chris 2009). The main content provided by the product-oriented sites is related to product characteristics, e.g. brochures, online transactions, and various enquiries and demonstrations. Information about the company, markets, culture and directions of development are integral to the corporate-oriented websites. According to Rayport and Jaworski (2004), good website design should address several interface design elements, which are important for the customers, such as context, content, community, customisation, communication, commerce and connection to other linked sites. It is not possible to satisfy personal expectations of each user but customisation can bring it closer to this goal. Accordingly, site customisation has started at high-profile websites, such as Google's regional settings and language changes according to the country of access. Design elements can be observed and applied in various ways, depending on the website's purpose and the targeted groups of users. For instance, site context can be aesthetically dominant, functionally dominant or integrated; content can be set up as product-dominant, information-dominant or service-dominant (Rayport and Jaworski 2004). Some authors like Oxley and Miller (2000) showed that the relevance of the content and the design of the presentation have very strong influence on the perceived quality of the website.

Information content undoubtedly has the utmost importance for the user. Rayport and Jaworski (2004) compared Internet technologies to the traditional media from the perspective of information content, and they identifies eight significant differences. Rational product-based information characterises websites, while emotional brand values are better presented by traditional media. Websites are better for short-term, product-oriented brand-action goals and long-term corporate identity objectives, for prompt customer action and the cost is not related to the size of the target audience. On the other hand, traditional media strongly builds long-term values, awareness and emotional brand values. Among these differences, probably the most prominent one is that cognition dominates website use, compared to emotion in the case of traditional media (Rayport and Jaworski 2004). An additional issue is the rapid change and frequent emergence of new ICT solutions that support a website, making it hard for website owners to stay attuned with the best technological options regarding website speed, response, security of data etc.

Small companies are generally cautious about comprehensive adoption of Internet solutions, such as business-to-business technologies mainly because they relate them to high investments with uncertainty of any increased profit (Zheng et al. 2004). There are many models of how to effectively adopt ICT as a business support (Hameed et al. 2012; Redoli et al. 2008; Zheng et al. 2004; Ngai 2003; Davis 1989). There are many studies related to the impact of Internet technologies on SMEs in various countries, such as in Ireland (Doherty et al. 2016), Spain (Soto-Acosta et al. 2016; Lopez-Nicolas and Soto-Acosta 2010), Belgium (Caniëls et al. 2015), the UK (Wang and Ahmed 2009), Scotland (Galloway et al. 2011), Austria (Haas 2002), India (Javalgi et al. 2012) and Taiwan (Ramanathan et al. 2012). However, there is still a need to investigate the factors that influence decision making in the process of website adoption. One of the theories that are very often used as a basis for analyses of various factors that affect the adoption of innovation is the perceived-risk theory (PRT) by Roselius (1971), from the field of social psychology, and it emphasizes that adopters of innovation must evaluate the potential benefits against possible risks. Corporate websites can be viewed as innovative marketing tools. Flanagin (2000) elaborated that the organizational characteristics, perceived benefits of innovations and social pressures are three main factors influencing the adoption behavior of the organisation. Other researchers adopted a slightly different model, emphasizing perceived benefits, organizational compatibility, technical compatibility, complexity and top-management support as the major influential factors (Beatty et al. 2001).

Davis (1989) developed a widely accepted model for the use of new technologies, based on two specific variables, perceived usefulness and perceived ease of use, as fundamental determinants of user acceptance. There are different levels of adoption of new internet-related technologies, from basic email use, then pure web presence without any dynamics, further progressing to prospecting and business integration and up to business transformation at the highest level (Bingley and Burgess 2012; Teo and Pian 2004). Investigations of relationships between the type of business and website adoption showed that the market sector is a very significant factor, as well as adequate infrastructure and adopted Internet strategy (Doherty et al. 2016).

For companies, especially for the young ones and SMEs, the benefits of website deployment are hard to perceive at the very start, due to many factors, such as access and use, security and the unproven commercial benefits (Flanagin 2000). Adoption of a corporate website requires investments in necessary supporting hardware, software and website development, as well as employee training and usually external services of website hosting. Managers need to justify these costs with tangible benefits. Flanagin (2000) set up a model of organizational-website adoption emphasizing three main success factors: organisational features, perceived benefits and social pressures. He elaborated that social pressures on the inter-organizational level critically determine adoption of innovation within an organisation. His studies of the organisational size aspects showed that large companies adopt innovation earlier than the small ones, mainly due to the availability of necessary technological and human resources. Organisational age rather negatively affects the innovative behaviour, indicating that those with greater reliance on advanced technologies more readily adopt innovations. Introduction of innovation to the regular business activities is related to certain risks and always results in increased expenditure, consequently necessitating perceived direct benefits in order to justify innovation.

Generally, websites are perceived to increase communication with customers and the information flow. On the other hand, websites that are highly technically complex web portals can inhibit adoption and/or communication. In the case of a high level of ambiguity, inter-organisational social pressures have been recognised as a very influential factor, especially in the case of micro-SMEs (Flanagin 2000). For instance, ambiguity is present when the company has no ICT department or does not employ ICT experts, which often is the case for micro-SMEs. In such cases, if an organisation sees that others in its business field have a website, it is highly probable that it will start one also. Organisations that perceive themselves as leaders in a field are especially prone to social pressure. However, unavailability of proper information or lack of knowledge about applied technology might result in thinking that website adoption is not particularly important.

Some benefits might not be immediately and directly seen. Weltevreden and Boschma (2008) showed that important indirect benefits exist, such as improvement of performance through improved relationship with customers, an increase of customers and growth of sales, as a result of an effective Internet strategy. Some recent studies undoubtedly have proven that adopting a website can produce increased profits, as shown by Scaglione et al. (2009). Websites offer substantial opportunities to achieve potential benefits for a company, including online promotion. Frequent promotion techniques used on a company website include blogs, advice columns, or frequently asked questions (FAQ), discussion groups, virtual tours, live events (e.g. webinars), various affiliate programs, free articles, e-books and brochures, voting polls and more. On the other hand, for the website to succeed, it is important to promote it through search directories, keywords advertising and affiliate marketing.

The main objective of this study is to explore the factors that influence website adoption by the smaller companies that mainly do not have their own ICT departments. The study investigated the attitudes of SME managers towards Internet-mediated business development and promotion, comprising a case study of SMEs in Serbia. Special focus was given to low- income micro-SMEs. Innovation is an essential foundation for market competitiveness, but adoption of innovation relates to very complex questions because it is influenced by many factors, and successful models are subjects of current research in various fields (Ordanini et al. 2014; Hertel and Menrad 2016; Hosseini et al. 2016; Soto-Acosta et al. 2014). Adoption of a website as innovative tool for marketing and customer- relationship management is presented here as one simple exemplary case to highlight some influential elements governing the adoption of innovation in companies. In general, websites have existed for a long time now, and, from that perspective, it is hard to characterise them as innovative tools. However, many new web technologies have emerged in recent years, especially in business dealings, and adoption of these innovative tools can efficiently reflect general attitudes towards adoption of innovation in small companies. The available budget for innovation in small companies is usually very low, especially in Serbia. Consequently, it is hard to collect and analyze a proper number of case studies in innovation areas requiring high investments, whereas innovative Internet technologies are related to relatively lower investments, if compared to new production technologies. An additional rationale behind the selection of website adoption as our case study is the fact that nowadays almost all companies, regardless of their size, have a company website. The level of a company's website development reliably reflects the level of adopted innovative tools in this area, from basic websites to innovative customer-relationship management dynamic systems. This research is designed as an initial study to provide basic insights into the attitudes of company owners toward innovation and to identify those factors that would support innovation adoption in small companies in Serbia. Further investigation is planned to consider a wider set of factors, based on these preliminary studies.

2 Method

This research is based on the case-study approach. The companies involved in this study wished to remain anonymous. The data for this research was collected over the period of six months. Within the study, data was collected through structured questionnaires, semi-structured interviews with owners and general managers of small enterprises (who are also the marketing managers) or with marketing managers in medium enterprises. Documentary evidence, such as printed brochures, corporate reports, company websites and business news, was also considered, providing additional information. After the preliminary analysis of the answered questionnaires, meetings with SMEs representatives were held again to review and discuss the data, in order to get general insights into the differences between adopters and non-adopters of a company website, regarding their managerial attitudes. These interviews were guided by, and included, the main theme and several related issues: (1) What factors influence a company to have a website? (2) What factors influence a company to use its website as a marketing tool? (3) What influences are dominant in those companies which do not adopt a website? (4) What are management's attitudes about using a website as a communication tool with customers? (5) Is there ambiguity surrounding adoption of innovation in general and website as innovative technology in business?

This method, which combines both quantitative and qualitative research methods, was selected because it is particularly well suited to the study of dynamic markets (Haas 2002). It is based on a mixed-methods research model (Burke and Onwuegbuzie 2004). Such a case-study approach is appropriate when the existing research data are rare or do not exist. It enabled a collection of rich and detailed information and better understanding and interpretation of quantitative results obtained from the survey questionnaires (Gable 1994).

3 Results

The research was based on a study of 32 small and medium enterprises (SMEs) in Serbia. The majority of interviewed companies were in the category of micro-enterprises (with less than ten employees). The distribution of enterprises according to the number of employees is given in Table 1.

The results have showed that four out of 32 SMEs (or 13%) do not use the Internet in their daily business operations and communication (Fig. 1a), while 12 SMEs (or 38%) use only email communication and do not have their own website (Fig. 1b). Twenty companies (or 62%) have corporate websites. We define adopters of the website as those that deployed their own company website and non-adopters as those that do not have a website, but they still can use some form of online promotion by using external online services (e.g. banners, Facebook, etc.).

Most of the surveyed SMEs expressed that the company website is significant for business development, but they rank its importance in various degrees (Fig. 2). From Fig. 2, it can be seen that the majority of investigated companies (30 SMEs or 93.75%) consider the company website as significant for business. However, two companies perceived a website as insignificant for business development and promotion. These two companies had locally oriented business and very low budgets for marketing.

Online promotion done by using external online portals not their own website, such as banners, link exchange, etc., was used by 21 SMEs (or 66%). Respondents were offered to select among five categories of online advertising, giving them the opportunity to choose several answers. The category of online advertising "other forms" was most often selected (14 SMEs), followed by "banners" (13 SMEs), as given in Fig. 3. Category "other forms" comprised social networking sites (Facebook, MySpace, etc.), affiliate marketing networks, video exchange sites (YouTube, etc.), blogs (corporate blogs or other), in-game advertising and other.

The majority of the surveyed SMEs (30 SMEs) appreciated online promotion, but two of them did not think it effective at all (these two companies did not have their own website and did not plan to adopt one) (Fig. 4). Seven SMEs experienced a "great" effects by using the categories of online advertising given in Fig. 3.

Figure 5 shows how SMEs ranked five different roles of a company website in business development and promotion. Respondents were asked to assign a number of points (1, 2, 3, 4 or 5) to each of provided answers. A higher number of points

Table 1 The distribution of surveyed SMEs according to the number of employees

Number of employees	Number of surveyed SMEs
1–10	24
11–50	2
51–100	3
More than 100	3

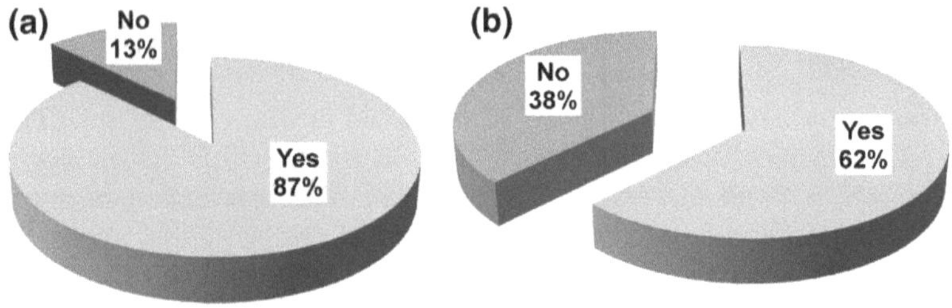

Fig. 1 Percentage distribution of answers to the following questions: **a** Do you use the Internet in daily business operations and communications? **b** Does your company have a website?

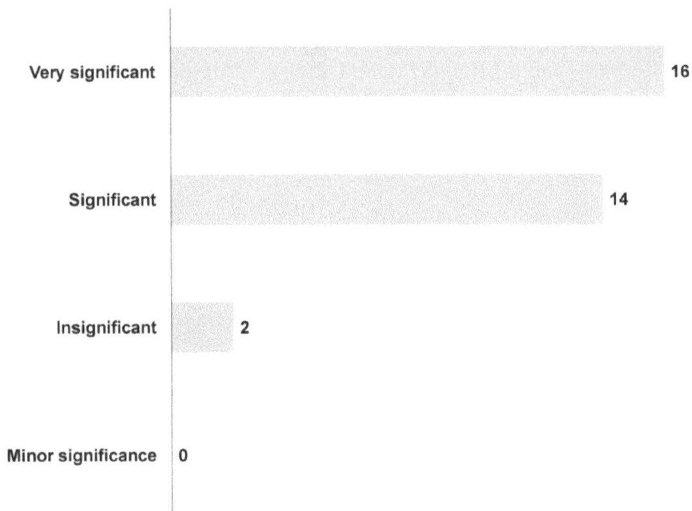

Fig. 2 Number of SMEs who selected one of the four options to the following question: How significant is a company website for business development and promotion?

corresponds to a higher significance level, thus the answer is higher ranked if it obtained a higher total sum of points. It is obvious that both adopters and non-adopters considered that the most important role of the company website is to present the public image of the company and to create positive impressions among the stakeholders.

If we further examine only the highest points assigned to questions in Fig. 5, it can be seen that the distribution is similar, as given in Fig. 6. Figure 6 shows the distribution of maximum assigned points, whereas Fig. 5 shows total sums of points, per question. It can be clearly seen that the positive impression that a website produces was valued twice as high above the feedback it provides.

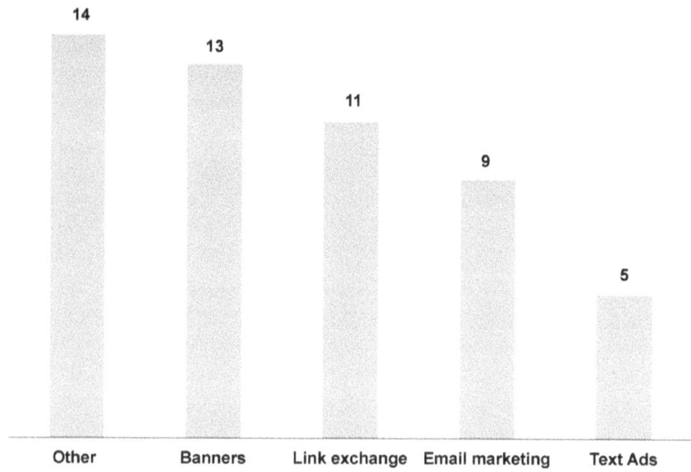

Fig. 3 Number of SMEs who selected the indicated answer to the following question (multiple choices permitted): Which promotion categories do you use on external online portals, excluding your own website?

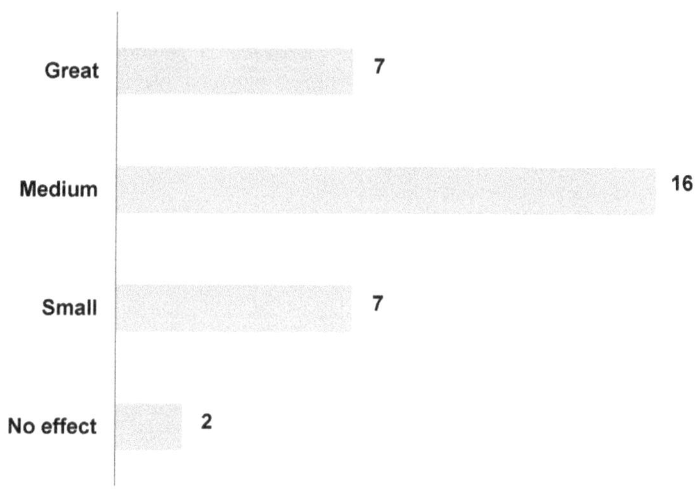

Fig. 4 Number of SMEs who selected one of the offered answers to the following question: How would you describe the effect of online promotion on your business, per categories given in Fig. 3?

Also, Fig. 6 indicates that non-adopters generally do not recognise the opportunities offered by a website to dynamically interact with customers, business partners, public and other important stakeholders of the company.

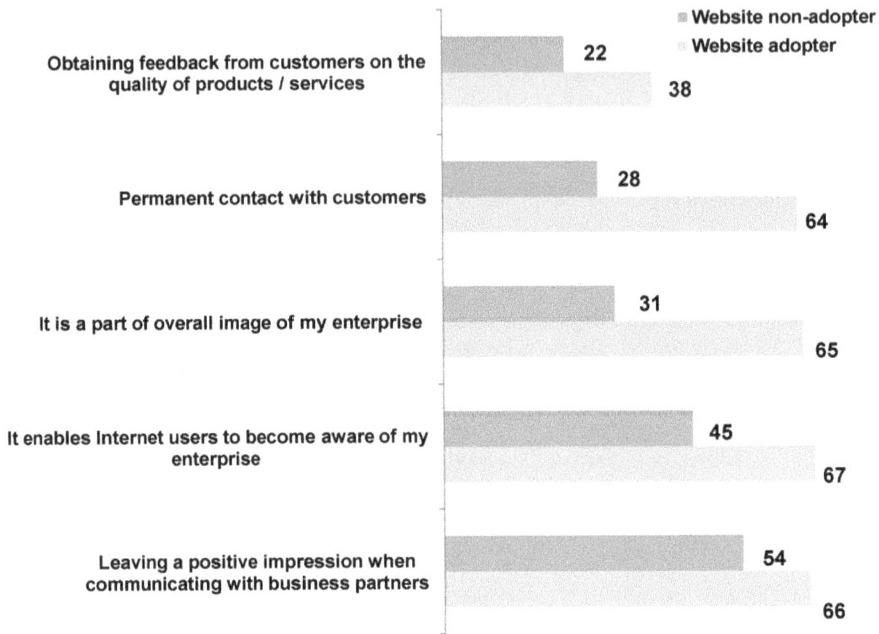

Fig. 5 Five different tasks of the company website, according to the number of assigned points, by responses to the following question: What is the most important role of the company website?

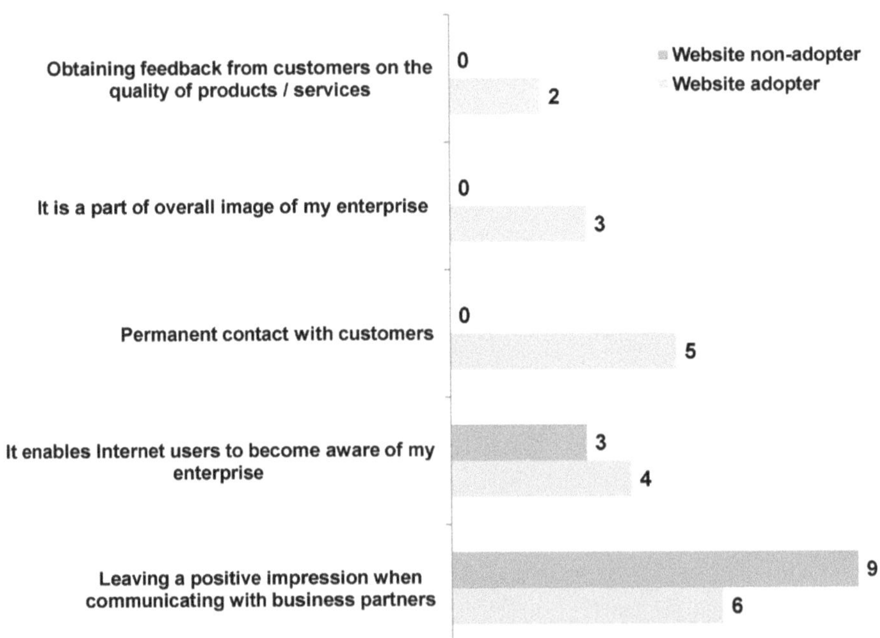

Fig. 6 Number of SMEs who assigned the maximum points (5) to one of the offered answers to the question: What is the most important role of the company website?

4 Discussion and Implications

The results showed (Fig. 1a) that a rather high proportion of SMEs (13%) does not use the advantages of the Internet at all. Also, 38% of all investigated companies identify only e-mail communication (Fig. 1b) as the basic form of the Internet use. Only 62% of correspondents have their own website. On the other hand, the majority of surveyed SMEs (30 SMEs or 94%) recognised the importance of a website (Fig. 2), but at the same time they mostly do not recognize the need to significantly invest in the development, regular maintenance and promotion of their websites. Only a few companies have staff engaged in website maintenance, while most of them considered that once they had started the site, the online promotion is done. There is a lack of knowledge about the comprehensive role of the website, together with a lack of proper communication with website developers resulting in low quality websites, from aspects of design, speed and reliability, and especially from aspects of interactivity, navigability and content. The interviews with managers indicated that, especially, micro-SMEs do not recognise the need for a change from a product-oriented business strategy to the customer-focused strategy. Two of the surveyed companies even denied any benefits of a corporate website and perceived web technology only as an unnecessary expense related to "fashionable novelty" (Figs. 2, 4).

Results presented in Fig. 3 showed that link exchange and e-mail marketing were often used as forms of advertising. And 23 out of 32 SMEs indicated that other forms of online promotion, besides the website, have influenced their business (Fig. 4). Here we assumed that answers "great" and "moderate" together constituted a major influence. Through conversations with respondents, we found out that they use e-mail marketing in a way that resembles the print- advertising model. Although there are many similarities, it is important for SMEs to acknowledge that e-mails are much more complex than adverts in printed media (Brock et al. 2016; Ellis-Chadwick and Doherty 2012). E-mail correspondence can be a very effective tool to determine various issues related to the company's business, as presented by Haas (2002) who found that analysis of customer complaints sent by e-mail had helped to discover and resolve problems associated with products and services. Analysis based on personal interviews also showed that SMEs in this research considered these other forms of online promotion (e.g. banners) as only static advertisements. Several researchers showed that these are complex and dynamic forms of advertising, and, although design and visual effect are very important, banner recognition primarily depends on understanding the users' affiliations and behavior (Laffey 2007; Calisir and Karaali 2008; Hamborg et al. 2012; Syrkiewicz-Switala et al. 2016; Li and Du 2017).

The selection of websites for online advertising is very important and has been studied by other researchers (Ngai 2003), and our results confirmed it through observation of various promotional categories on online portals (Fig. 3). Also, selection of the proper keywords is of the utmost importance, for both the website and other forms of online promotion, and should be done with great care, taking into account the target groups and attitude of the customers rather than the firm managers or employees (Kim et al. 2012; Anupam et al. 1999). Semi-interviews helped us to come to the conclusion that by "other forms" of promotion most of the surveyed SMEs meant social-networking sites, especially Facebook, because they can create a profile and serve as advertising, with zero budgets. Personal contacts and observation of the activities of surveyed companies on Facebook showed that they have not yet adopted a proper approach to that form of online promotion. Promotion on social-networking sites requires a strategic approach as does any other marketing media, and it should be considered as a modern tool to achieve traditional objectives in marketing (Weinberg and Pehlivan 2011; Hohmeier and Borja-Hart 2016).

Results presented in Fig. 5 have shown that the answer "leaving a positive impression when communicating with business partners" garnered the highest total score (120 points), twice the total score of the answer "obtaining feedback from customers on the quality of products/services". It is obvious that the website role was primarily recognised as a means to create an impressive image of the company for random visitors and company target groups. Results given in Fig. 6 showed that 15 SMEs (around 47%) considered "leaving a positive impression in communication with business partners" the most important role of their website, giving this answer the maximum points. Only a small number of surveyed firms recognized that the feedback from customers and interactive communication with website users (customers, employees, prospect customers, business partners and company stakeholders in general) is a very important aspect of the website's existence. The conducted interviews also proved that the majority of surveyed companies had not adopted a clear approach to the use of an existing website. They mainly conceived of a website as simply advertising and did not recognise all the benefits it could provide.

Results in Fig. 6 showed that seven SMEs perceived that the most important website role is to enable Internet users to become aware of the company, while only two SMEs used all types of promotion cited in the survey. Among the twelve companies that did not use any form of online promotion, there were eight without a website and four companies with their own website (Fig. 1). The fact that they did not use any form of advertising makes their websites "invisible" to the Internet users. Those companies (with so-called invisible sites) were all among the 15 companies that considered making a good impression on business partners the most important role of the website. Later adopters or non-adopters of a website shared a rather common belief that a website is necessary only in order to maintain their current position. Also, in many cases, organisational support was lacking because technological infrastructure in these companies was not fully or completely compatible with the web technologies. Beatty et al. (2001) surveyed 286 companies and found significant differences among their reasons for the adoption of the corporate website. According to their investigation, early adopters emphasized perceived benefits and compatibility with their existing technological infrastructure, as well as with organisational framework, which in return generated organisational support for adopting a website.

It is obvious from Figs. 5 and 6 that the majority of interviewed companies perceived a website mainly as the tool for making a good impression. This implies that the aesthetic dimension of the website is mainly exploited while other characteristics, such as interactivity, navigability and content, besides design, are mainly not perceived at all. Karayanni and Baltas (2003) noted these as the four main characteristics of a successful website. Analysis of attitudes of specific users of a corporate website should be realised prior to website development. Herrero and Martín (2012) showed that, in the case of tourism accommodations, acceptance of the services offered by a website is determined by the usefulness and the ease of use of the website, as perceived by users. Hernández et al. (2009) discovered that users need navigability to feel secure and comfortable, but also that the information provided on the website must be accurate, informative, updated and relevant to customers' requirements. Non-adopters perceive the company website mainly as a tool to present nice images and simple, static information about the company to the general public. Very often, they do not see their business developing beyond the local environment, consequently thinking that it is enough to use traditional means of promotion such as brochures, leaflets and local newspapers ads. The strong impression gained during our live discussions with managers of non-adopters indicated that they generally had a low level of knowledge about new ICT advancements, especially those related to website deployment for business development. They knew almost nothing about possibilities to dynamically interact with website users and had only the most basic perception about its possibilities, such as plain, static presentation of images and texts. Our results are in accordance with the results of Bortamuly and Goswami (2014), who investigated the sector of the handloom industry and showed that education of owners is crucial for adoption of modern technology in industry, especially in the case of small household-based companies.

A well-recognised characteristic of a corporate website is its ability to deliver clear and precise messages to a targeted audience, such as comprehensive product information, services details, designs, images, video and audio product/services descriptions, technical drawings or various manuals and various other customized info to meet the needs of customers. Our study showed that the management of surveyed companies is aware of these possibilities to a certain extent, and it seems that other issues are more likely preventing dynamic use of existing corporate websites. New ICT technologies enable web developers to deploy customized interactivity, multichannel marketing, personalisation, mobility and speed, along with efficiency and enhanced relationships with target groups. Besides efficiency and precision, emotional satisfaction is related to repeated visits to a website, and design characteristics are essential to the length of stay, and branding starts to be a very important part of web presence (Chris 2009). Our study showed that only two companies (Fig. 6) having a website recognised that new technology can help them to create targeted messages to specific groups, and others mainly focused on emotional satisfaction based only on design. However, even the design of a website was mainly representing the image of the manager and/or owner, who very often had no educational background in any design-related field. In many cases, webpage design is done according to the wishes of the company owner, even without consulting professional designers or ICT experts, and very often results in a very low-quality website from the perspectives of design, color, images, navigability and speed. This, on other hand, inevitably results in very few website visits, which is in accordance with research done by many others (Alkindi and Bouazza 2010; Belanche et al. 2012; Calisir and Karaali 2008; De Marsico and Levialdi 2004; Fang and Holsapple 2007; Hasan and Abuelrub 2011; Kim et al. 2009).

Mehrtens et al. (2001) deduced that three factors significantly influence the adoption of the Internet by small companies: perceived benefits, organisational readiness and external pressure. Wang and Ahmed (2009) emphasized two factors as primarily determining Internet technology adoption, namely, "external pressure" and "perceived benefits." Results of our study clearly indicated that Internet adoption by interviewed SMEs can be considered as an introduction of innovative strategy, since there was a rather large lack of knowledge and skills related to Internet technologies. This, in turn, generated a large percentage of small firms that completely ignore the advantages of modern communication technologies (Fig. 1a) and for others to engage in only a limited slice of the possibilities (Fig. 1b). McCole and Ramsey (2005) found that six factors are related to adoption or non-adoption of Internet technologies in a professional service firms: opportunities enabled by technology; perceived barriers; technology capability; response to innovation; customer orientation; and sensitivity to customer environments. There are different models for the assessment and development of Internet-based information and communication services in small and medium enterprises.

(Hosseini et al. 2016; Redoli et al. 2008; Zheng et al. 2004) by which a company can overcome the gap between understanding of the business process and lack of knowledge about technology, or vice versa.

Based on this study, companies in Serbia mainly use their websites for the projection of the company image and only a few of them include systems such as Customer Relationship Management (CRM), Business-to-Business (B2B) or Business-to-Consumer (B2C). These few are usually companies related to the sectors of banking, insurance and/or telecommunication. This is especially true in the case of micro-SMEs with locally oriented business. Just recently, a small number of corporate websites appeared that completely support e-sales, and those are mainly limited to computer equipment. Analysis of the Internet-marketing use by European companies showed that most of it is related to e-business or sales strategy, and a large number of companies emphasised the importance of e-procurement (Tiago and Tiago 2012), as well as implementation of CRM (Alshawi et al. 2011). According to the literature, several factors affect the adoption of e-commerce web technology by companies in which functionalities and the cost of the website are identified as the most influential factors (Hong and Zhu 2006).

Beatty et al. (2001) studied the differences between the reasons for website adoption, and they also emphasized perceived benefits and compatibility with existing technology as determining factors. However, we see both of these factors strongly conditioned by the level of knowledge about ICT possibilities and Internet technologies. Our results also showed that the size of an organisation strongly influences the core skills or knowledge, in general. Small companies have adequate knowledge about website advantages only in the cases when they are in an ICT business or some of the employees have an educational background in ICT. This was confirmed by the attitude of two companies interviewed in our study, who both had a strong resistance to adopting a website or using any Internet technologies, in general. These companies both have less than ten employees and owners who solely make the business decisions or have no educational background or no ICT skills of any level.

Triandini et al. (2017) studied the influencing factors on corporate website adoption and developed further a model proposed by McCole and Ramsey (2005). The most important reasons to adopt a website include: (1) promotion and advertising the company's name and mission; (2) communication about specific product/service; (3) enhancement of customer service; (4) communication with customers and suppliers; and (5) to imitate and/or to be better than competitors. Triandini et al. (2017) developed the model and included many more aspects related to 13 benefits, 51 functionalities and 9 non-functionalities that e-commerce should provide to the company. The main perceived benefits should comprise effective advertising and brand building; increased number of customers; increased sales, profits and cost savings; and establishing new suppliers and better purchasing terms. McCole and Ramsey (2005) stated that some of the most important internal environment factors influencing the company decision to develop a corporate website are: management resistance; technology concerns; resource issues; lack of awareness; and lack of information, and these factors remain the most significant even today. Our investigation strongly supports their conclusions, with the addition of market-oriented behaviour including countering competitors' actions. It is a common fact that technologies do not enter industry easily. Investigations related to the technology diffusion in industry revealed that the lack of information about new technology, as well as company's goals and vision, has great influence (McCole and Ramsey 2005; Triandini et al. 2017).

However, it is also important to properly evaluate the necessary level of Internet engagement in company business. Some SMEs are mainly locally oriented, and extensive investments of the resources in comprehensive Internet solutions might be unproductive. For instance, Galloway et al. (2011) showed that, in the case of small rural firms, the best solution was to foster a local orientation of the portal, in order to facilitate local trade, rather than adopting an external market orientation. The model of Internet adoption should take into account the specific difference between SMEs and large firms (Zheng et al. 2004). Our results support the findings of others (Flanagin 2000; Galloway et al. 2011; Weltevreden and Boschma 2008; Haas 2002; Wang and Ahmed 2009; Zheng et al. 2004; Doherty et al. 2016) who also stated that the type of business is one of the determining factors for the way the company develops strategy towards Internet use. Weltevreden and Boschma (2008) suggested that a properly adopted Internet strategy can positively influence business performance from the perspectives of more customers, improved customer relationships, sales growth and competitive advantage. Our study showed that company strategy towards Internet sales is more beneficial for those having regional and wider target groups of customers than for those with locally oriented business. Our findings support results presented by Konings and Roodhooft (2002), based on comprehensive analysis in Belgium firms, that e-business has no effect on the total amount of productivity in small companies, but they found positive performance effects related to large companies. However, research results suggest that websites adopted for small companies might produce a positive impact on their performance, possibly by improved marketing and distribution (Konings and Roodhooft 2002; Scaglione et al. 2009). Only two companies we interviewed had adopted an Internet strategy, and two companies had no plans or strategy towards it. We can speculate that the two companies that had no plans to adopt any Internet related activities are very small household-based companies and their owners consider a website and Internet use solely as a mean to broaden their business, which they do not plan. Additionally, any Internet-related activity for them was suffused mainly with uncertainty.

The nature of the product offering and competitors' behaviour are important factors in adoption or non-adoption of a website. In the case of larger companies, they adopted a website when their competitors seemed to enjoy increased benefits, especially in the depth of product offerings for various types of customers. However, in the case of very small companies, our interviews indicated that competitors' behaviour had no influence on website adoption, which accords with findings by Aboelmaged (2014). But the prominent factor noticed during our analysis of the questionnaires and live discussions was the ambiguity usually surrounding the level of development of technology solutions that determines ease of use for those who do not understand the core essence of technology behind the website. The ICT world is rapidly changing and following newly emerging technological solutions is not easy, even for those with an official ICT background. It demands significant investment of time and resources to know the latest developments in the ICT area, especially in regard to aspects of data security and the reliability of applied solutions. Altogether, within our study, these factors result in slow websites and a low quality of design, especially as regards navigability and interactivity. Grant et al. (2014) explored the risk perceptions of key stakeholders in SMEs when making decisions on technology investments. They found that subjective elements are important in the assessment of technology risks and that the role of the company's decision makers is crucial in risk perception. Hosseini et al. (2016) showed that a significant barrier to new technology adoption is related to the high risk perceived by managers on return on investments (ROI).

Another conclusion we can make based on this investigation and especially based on live discussions is that, generally, companies tend to consider the target groups of the website users as distinct from the real company stakeholders. Regular updating of information is of the utmost importance from aspects of repeated visits of users. Completed interviews revealed that only a few managers recognised this and its implications for their customers. The website was primarily considered as a means of communication with younger communities and potential future customers. Discussions revealed that almost all surveyed companies in this study mainly use traditional methods for interacting with investors and business partners, such as meetings or phone conversations, and do not consider a website or email to be able to provide an efficient tool for this type of communication. They all expressed concerns about security of confidential data within the online environment, data that they exchange with their business partners exclusively by traditional post or at meetings, which is particularly important when managers are asked to justify investment in online activities. Individual characteristics of each company must be considered as an important precondition of the decision process.

All interviews conducted within the scope of this research indicated that security risks related to the published information on their website are considered as very serious issue for their business. Various areas of applied research are related to different confidential data concerns. The authors of this paper have been engaged in various applied research activities and commercialization of those results in cooperation with the various companies in the region and have conducted interviews with company owners. We found that, in space research and new nanomaterials, the stakeholders demanded a high level of security for their data input (Matovic et al. 2016). This was also the case for medical research for commercialization of new materials are used by high profit companies very unwilling to disclose the research details (Zivic et al. 2013); and new technologies applied in furniture industries that decrease cost production are very competitive and thus demand data security (Grujovic et al. 2016). Even the traditional industries involving upgrades of processing technologies or product improvement (Mitrovic et al. 2014; Vencl et al. 2013; Babic et al. 2010) are not willing to share data openly. For all these very diverse companies that applied research for innovation, the common issue with company websites has been data security. Our interviews indicated that the majority of these high innovation companies either strongly doubt the efficacy of the existing website security solutions for dynamic content safety or lack awareness of existing options related to protection of the website content from public access. Accordingly, they mainly envisioned only static website content including rather simple information.

Ordanini et al. (2014) showed that customer adoption of new services is very complex, and they elaborated that adoption of innovation is strongly influenced by complex attributes and their collective influences. They also showed that the better approach in service adoption is to consider several alternatives in influential attributes than only one, even if it is currently perceived as the best one. The conceptual framework drawn by Ordanini et al. (2014) recognises the core attributes in new service adoption to be: novelty, complexity, meaningfulness and relative advantage, additionally influenced by coproduction requirements (high, medium and low), whereas they proposed adding fit logic and configuration theory. Supporting theories are proposed also by Woodside (2014, 2015, 2013), who elaborates that complex relationships exist between recognised influential parameters that determine the system behaviour and that modeling multiple realities is necessary. He introduced complexity theory, contrarian case analysis, asymmetric tests versus symmetric thinking and configural analysis and demonstrated that interdependencies between influential parameters should not be considered as the simple consequence of the sole dependence between two parameters only, but also conditioned by many other system properties that determine the direction of the system development over time. A very good case study of complexity theory's application to the service improvement (or adoption of innovations in services) and the influence of contrarian cases is given in Wu et al. (2014).

Different types of sites have different characteristics to be evaluated. Many authors have studied website quality and its impact on various types of companies (Kim and Niehm 2009; Kim and Stoel 2004; De Marsico and Levialdi 2004; Webb and Webb 2004; Lombard and Snyder-Duch 2001). Quality is an attribute of a product/service that indicates how well it meets consumers' needs (Kim and Niehm 2009). Kim and Niehm (2009) stated that some website quality dimensions were somewhat more important than others, such as interactivity, online completeness, ease of use and entertainment. They found that online completeness and entertainment greatly influence customer loyalty, taking into consideration the relevancy of information provided by the website. Many researchers have investigated ways to collect data on the behaviour of the site users (Bucklina and Sismeirob 2009; Luo et al. 2011; Pakkala et al. 2012). Further research on aspects of the website quality and Internet-based marketing for small firms, involving a larger number of SMEs in this region, would reveal if all of the previously listed factors are equally important, when considering specific characteristics of SMEs.

Lee and Kozar (2006) presented useful insights for the development of high quality websites. They found that the highest quality websites resulted in the highest business performance. It is necessary for decision makers in a company to gain a better understanding of online customers' perceptions of desirable websites (Lee and Kozar 2006). This is in accord with the results of our study, showing that decision makers in micro-SMEs are mainly owners, and the definition of the website quality is closely related to their perceptions, not necessarily considering users' attitudes. This should be changed to the customer perceptions instead (Alkindi and Bouazza 2010; Belanche et al. 2012; Rosen and Purinton 2004). However, general users' perceptions are hard to define because the characteristics of all users are hardly foreseeable (Vila and Kuster 2011; De Marsico and Levialdi 2004). Vila and Kuster (2011) have analyzed the reaction of 350 visitors to the properly designed website of a fictitious clothes company, created by experts. Then they offered the visitors several modified versions of the previous website, created by removing some of the main features (web security; customer service; amount and quality of information provided; and usability). Visitors satisfaction was measured by using 21 indicators, and the results only proved that even the best professionally designed websites cannot satisfy customers for all 21 indicators.

Classification of websites can be done according to their function as: (1) online storefronts; (2) Internet presence; (3) content sites; (4) electronic malls; (5) incentive sites; and (6) search agents (Teo and Pian 2004). Another classification can be done according to the business purpose as: (1) promotion of product and services; (2) provision of data and information; and (3) processing of business transactions (Ho 1997). Websites can be viewed in various ways, depending on the mission and vision of the company and the role they perceive for it. Versatile functions that a website can perform are investigated by many researchers, each providing insights into a specific characteristic. However, our study showed that, in the case of micro-SMEs which are not doing business in some ICT area, websites are mainly considered as a marketing method. Also, a majority of managers emphasized that a website is only an additional expense, not perceiving it as an investment into better marketing. This is in accordance with results presented by Whitehead (2008), who stated that marketing communication is often considered as a variable, which can be discarded in times of recession. Having in mind the very low budgets available at investigated micro-SMEs in our study, it is obvious that such an attitude can be expected.

The majority of companies that we interviewed emphasized quick return on investment and tangible benefits as the main expectations from a website. This is, however, a rather complex matter to be precisely quantified due to many factors. Perceived return on the investment of time and resources is hard to evaluate, especially in the case of the online marketing role of a website. Scaglione et al. (2009) conducted a study related to the relationship between the generated profit and an adopted website with a focus in the hospitality industry, and their results confirmed that, in this sector, hotel revenue was higher after website adoption and that revenue growth rates were higher for those with their own website than for hotels with only a web presence at regional portals. There have been various studies related to the measurement of online visibility. Drèze and Zufryden (2004) developed the visibility measure to capture the extent to which a user will probably come across a reference to the website of the company within the vast extent of the Internet resources. They based their model on public directory listings, website contents and the most influential search engine results (Drèze and Zufryden 2004). Many factors influence the website visibility, such as online advertising, links with other websites, search engines, news reports, chat rooms, social networks (e.g. Facebook), e-mail correspondence or integrated internet marketing but also offline means of website advertising and news reports (Tiago and Tiago 2012; Drèze and Zufryden 2004). There is a lack of comprehensive analysis and evidence related to the influence of the Internet use on the marketing activities and quantitative performance outcomes (Tiago and Tiago 2012). Online visibility of the company website is one of the key success factors and is strongly related to the website traffic as shown by Drèze and Zufryden (2004).

In this study, interviews with managers and company owners highlighted several issues related to the development of a corporate website. Micro-SMEs do not easily decide on the website structure and especially about investor-related sections. Business vision as a fundamental for adopted Internet strategy is important, but a significant number of micro-SME managers expressed ambiguity about what type of information is suitable for a website. It was obvious that they did not consider website structure as a variable that is connected to the target groups or even to the business-development directions. Mainly, this is a result of several factors affecting their everyday business, such as: lack of resources (low or no budget for innovation, low number of employees), lack of skilled personnel, unawareness of modern technology breakthroughs and lack of comprehensive knowledge on specific characteristics of proper online promotion. Lack of resources in a recognised barrier to circular business models (Rizos et al. 2016). Official statistical documents also show that a significant number of companies in Serbia has no Internet access and do not use or recognise the Internet as a business support (Statistical Office of the Republic of Serbia 2011). Among companies with Internet access, 20.7% received orders (excluding e-orders) in 2010, representing an increase of 0.7% compared to 2009 and 0.8% compared to 2008. Serbia has just recently started to use the PayPal system for handling transactions, and we did not include ordering and financial transactions in this study, choosing to stay focused on the most common types of Internet use in this region. Javalgi et al. (2012) reported a strong relationship between rapidly expanding businesses and Internet-enabled SMEs. A distinctive property of the Internet is its ability to make quick changes, especially if compared to traditional media (e.g. newspapers, TV).

Internet use and/or a corporate website can be seen as the introduction of innovation. Several stages within a process of innovation adoption can be established, and knowledge is recognised as a foundation for decision making and the first stage of the introduction of innovation (Chris 2009). When related to ICT and website technologies, knowledge comprises several interconnected parts, which influence perception: (1) professional knowledge and skills; (2) perceived ease-of-learning of new skills; and (3) perceived ease of use of deployed website technology. Our results indicate that lack of knowledge about ICT technology largely prevents its adoption and negatively influences recognition of possible benefits. In the case of micro-SMEs, the decision-making process is related mainly to the company owner, resulting in a company strategy closely linked only to his/her educational background. It will present a great obstacle to the adoption of any Internet or ICT related business activities when the owner lacks even awareness on new technologies. This was confirmed by our study, where, in the case of very small companies, all advanced stages of innovation adoption were essentially conditioned by the educational background of the owners, or their previously acquired knowledge.

Various Internet strategies, among which a website is a common tool to exploit Internet presence, can be adopted by various methods. Brochures provide limited product-and-services information, while catalogues can offer extensive and detailed data (Weltevreden and Boschma 2008). The help desk section of a website can provide users with various levels of services that are not fully enabled by the online catalogue, and proper deployment and maintenance of this part of a website can invariably positively influence customer relationships (Weltevreden and Boschma 2008). It was clear that, in the case of SMEs in this study, it is necessary to raise awareness of the company managers about the benefits of Internet innovation. Education on opportunities is needed, but also about the means to overcome barriers to aspects of social, technological and innovation. The introduction of various levels of Internet solutions (dynamic web portals, e-commerce, e-business, e-mail etc.) would strongly support successful growing of SMEs and their expansion into international markets, as well. Appropriate information management and continuous knowledge improvement are vital for the sustainable existence and development of the companies in a competitive market, nowadays. Most of the variables that affect website adoption are under control of the company management, and these should be carefully considered when aiming towards training in companies in order to raise awareness of the possibilities of a corporate website. However, a certain number of variables are uncontrollable, such as a company's available resources for innovation or some aspects of the external social environment, and these are very hard to influence. Knowledge is acquired through learning, and today's competitive market, where business is greatly influenced by Internet and ICT technologies, demands continuous learning. The decision to adopt, earlier or later, or to continue adoption of a website, online presence and interactions, depends on many factors, among which we find that knowledge is the most basic precondition and the most complex variable to determine rejection or adoption of a website. However, this was a simple study with a small number of interviews, and it should be significantly broadened in order to develop more general theories on how to efficiently introduce innovation in companies, or how to more closely link the business world with a world of university research.

5 Conclusions

This article has examined website adoption through a case study of small and medium enterprises (SMEs) in Serbia, with a special focus on micro-SMEs. The study examined company website presence as the tool for business development and the factors that affected website adoption. 62% of the surveyed SMEs have created their own website. The majority of correspondents recognised the importance of having a website, but with no clear comprehension of reasons, and they did not feel that the significant investment into the development, regular maintenance and promotion of their websites is needed. A large number of surveyed companies had not yet adopted a clear strategy for the use of the website. The most prominent role of the website was to enable the general public to be aware of the company's existence and to maintain the company's current position.

There was a lack of comprehensive knowledge about the possibilities of an innovative corporate website, due to the low level of knowledge about new ICT technologies in the cases of companies that did not have IT department and/or the owner had no educational background related to ICT. Accordingly, in these companies, there was a lack of organisational support to adopt a website, and the technological infrastructure was not compatible, as well. For a majority of interviewed companies, Internet and website adoption was perceived as the introduction of innovation. A large number of respondents adopted a website by following their competitors' behaviour.

There were differences in the approach of the companies towards website adoption and continuing application, depending on the type of business and the projected size of customers as perceived by the managers. Regional and more widely oriented companies recognised the benefits of Internet support aiming at a large customer population, whereas highly locally oriented companies did not see any benefits of a website, perceived it only as unnecessary expense and used only traditional methods of promotion within the local community. They were mainly household-based with a very low number of employees. Market-oriented behaviour and the nature of the product offering, including countering competitors' action, were very important for the decision process of a website adoption. Our study also showed that company strategy towards Internet sales is more beneficial for those having regional and more widely based target group of customers than it was for locally oriented business.

The knowledge about ICT possibilities influenced both the website adoption and the vision of the possible target groups—website users. In the case of micro-SMEs, the educational background of the owner essentially determines adoption, as well as the structure of the website and its use. Generally, companies considered website users to be different from their business stakeholders, and, especially, investors were not perceived as users. They viewed a website as aimed mainly at the younger generation and future customers. There was a distinct ambiguity surrounding the question of the appropriate website structure and type of data to present. Only a small number of investigated companies recognised that the feedback from customers and interactive communication with website users (customers, employees, future customers, business partners and company stakeholders, in general) are significantly important for business development. Altogether, this generates a gap in communication with site designers and developers, resulting in low quality websites, regarding the aspects of design, speed and reliability. Lack of education and appropriate technological infrastructure negatively influenced Internet use, corporate website adoption and awareness on any possible benefits.

The results of this study revealed that, in the case of larger companies with experts in ICT and marketing, the models proposed by Flanagin (2000) and Davis (1989) can be effectively used as guidance for website adoption. However, in the case of very small companies with limited budget and locally oriented markets, a few of those factors are more prominent, while some had very little influence on the adoption process. The educational background of the owner fundamentally determines adoption process, and all other determinants have very little or almost no influence. Another important influence is the annual income of the company, but, even with the available budget in micro-SMEs, owners will not consider introduction of this innovation if they have no knowledge of the background technology and cannot understand the risks related to it. This is in accordance with findings of other authors on innovation adoption (Hosseini et al. 2016). On the other hand, if the owner properly perceives the benefits of the website, the cost of website introduction and maintenance have no significant influence on the decision to adopt. Our results also proved that, for micro-SMEs, organisational support has no influence on website adoption, even in the case when other employees expressed strong support for website adoption. And while the owner's education is crucial for the decision to adopt, it has no direct influence on the extent of the website adoption, which is mainly determined by environmental and organisational contexts. In the case of micro-SMEs, social pressures can have an opposite influence to the case of larger companies. A significant number of owners resist following the competitors in the instance that they have no education about introduced innovation, and, not because they feel it is faddishness, but because they usually resist it until they are not familiar with the basic technology of its background, possibly indicating that specific cultural aspects of the region should be considered also as an influencing factor. But our results invariably showed that social pressures influence micro-SMEs to pursue further education in order to better understand innovations introduced by competitors. Results indicate that short workshops, seminars or online and distance learning can be efficient methods to overcome the knowledge gap produced by owners' age or formal educational background. Further research focused on the influences of cultural aspects and the interplay between factors on website adoption would provide more insights into the extent of its influence on innovation adoption, as well as on other website-adoption elements, such as perception of benefits and organisational and social aspects.

However, simple studies as this one can merely point to the complexity of the innovation adoption in organizations in general. Further studies, involving more parameters, a significantly larger number of interviews, and also including contrarian cases and new approaches such as application of complexity theory (Woodside 2013, 2014; Wu et al. 2014; Ordanini et al. 2014) should provide directions for efficient adoption models.

Acknowledgements The research study is part of project III41017 Virtual human osteoarticular system and its application in preclinical and clinical practice, funded by the Ministry of Education and Science of Republic of Serbia, for the period of 2011–2017.

References

Aboelmaged MG (2014) Predicting e-readiness at firm-level: an analysis of technological, organizational and environmental (TOE) effects on e-maintenance readiness in manufacturing firms. Int J Inf Manag 34(5):639–651

Alkindi S, Bouazza A (2010) An evaluation study of the navigation and search systems on two academic websites and Google. Int Inf Libr Rev 42:50–61

Alshawi S, Missi F, Irani Z (2011) Organizational, technical and data quality factors in CRM adoption—SMEs perspective. Ind Market Manag 40:376–383

Anupam V, Mayer A, Nissim K, Pinkas B, Reiter MK (1999) On the security of pay-perclick and other Web advertising schemes. Comput Netw 31:1091–1100

Babic M, Mitrovic S, Zivic F et al (2010) Wear behavior of composites based on ZA-27 alloy reinforced by Al_2O_3 particles under dry sliding condition. Tribol Lett 38:337–346

Beatty RC, Shim JP, Jones MC (2001) Factors influencing corporate website adoption: a time-based assessment. Inf Manag 38:337–354

Belanche D, Casalo LV, Guinalıu M (2012) Website usability, consumer satisfaction and the intention to use a website: the moderating effect of perceived risk. J Retail Consum Serv 19:124–132

Bingley S, Burgess S (2012) A case analysis of the adoption of internet applications by local sporting bodies in New Zealand. Int J Inf Manag 32(1):11–16

Bortamuly AB, Goswami K (2014) Determinants of the adoption of modern technology in the handloom industry in Assam. Technol Forecast Soc Change 90:400–409
Brock B, Carlson SC, Moilanen M et al (2016) Reaching consumers: how the tobacco industry uses email marketing. Prev Med Rep 4:103–106
Bucklina RE, Sismeirob C (2009) Click here for internet insight: advances in clickstream data analysis in marketing. J Interact Mark 23:35–48
Burke JR, Onwuegbuzie AJ (2004) Mixed methods research: a research paradigm whose time has come. Educ Res 33(7):14–26
Calisir F, Karaali D (2008) The impacts of banner location, banner content and navigation style on banner recognition. Comput Hum Behav 24:535–543
Caniëls MC, Lenaerts HK, Gelderman CJ (2015) Explaining the internet usage of SMEs: the impact of market orientation, behavioural norms, motivation and technology acceptance. Internet Res 25:358–377
Chris F (2009) Chapter 25 digital media. In: Marketing communications: interactivity, communities, and content, 5th ed, Pearson Education Limited, Prentice Hall, Essex, England
Davis FD (1989) Perceived usefulness, perceived ease of use and user acceptance of information technology. MIS Q 13(3):319–339
De Marsico M, Levialdi S (2004) Evaluating websites: exploiting user s expectations. Int J Hum-Comput Stud 60:381–416
Dimic M, Pavlovic A (2016) Actuality and perspectives of the wood industry development in Western Balkan countries. Int J Qual Res 10(1):131–142
Doherty E, Ramsey E, Harrigan P et al (2016) Impact of broadband internet technologies on business performance of Irish SMEs. Strateg Change 25:693–716
Drèze X, Zufryden F (2004) Measurement of online visibility and its impact on Internet traffic. J Interact Mark 18:20–37
Džupka P, Klasová S, Kováč V (2016) Analysis of innovative start-up companies: case of Košice region. Qual Innov Prosper 20:40–56
Eito-Brun R, Sicilia M-A (2016) An innovation activity model for very small entities in the software sector: an empirical study. R&D Manag. doi:10.1111/radm.12226
Ellis-Chadwick F, Doherty NF (2012) Web advertising: the role of e-mail marketing. J Bus Res 65(6):843–848
Estensen L, Ekambaram (2016) The role of competence brokering in creating industrial innovation network of SMEs in regional development. In: Abstract proceedings of the European conference on knowledge management ECKM, Ulster University, Northern Ireland, UK, September 2016, vol 2016, pp 242–250
Fang X, Holsapple CW (2007) An empirical study of website navigation structures' impacts on website usability. Decis Support Syst 43:476–491
Faustina F, Balaji T (2016) Evaluation of universities websites in Chennai city, India using analytical hierarchy process. In: International conference on electrical, electronics, and optimization techniques, ICEEOT 2016, DMI College of Engineering Palanchur, Chennai, Tamilnadu, India; 3–5 March 2016
Flanagin AJ (2000) Social pressures on organizational website adoption. Hum Commun Res 26(4):618–646
Gable GG (1994) Integrating case study and survey research methods: an example in information systems. Eur J Inf Syst 3(2):112–126
Galloway L, Sanders J, Deakins D (2011) Rural small firms use of the internet: from global to local. J Rural Stud 27:254–262
Gottfridsson P (2014) Different actors roles in small companies service innovation. J Serv Mark 28:547–557
Grant K, Edgar D, Sukumar A et al (2014) 'Risky business': perceptions of e-business risk by UK small and medium sized enterprises (SMEs). Int J Inf Manag 34(2):99–122
Grujovic N, Zivic F, Zivkovic M et al (2016) Custom design of furniture elements by fused filament fabrication. P I Mech Eng C-J Mec, Published online before print April 19, 2016, doi:10.1177/0954406216646401

Haas R (2002) The Austrian Country Market: a European case study on marketing regional products and services in a cyber mall. J Bus Res 55:637–646
Hamborg KC, Bruns M, Ollermann F et al (2012) The effect of banner animation on fixation behavior and recall performance in search tasks. Comput Hum Behav 28:576–582
Hameed MA, Counsell S, Swift S (2012) A conceptual model for the process of IT innovation adoption in organizations. J Eng Technol Manag 29(3):358–390
Hasan L, Abuelrub E (2011) Assessing the quality of websites. Appl Comput Inform 9:11–29
Hernández B, Jiménez J, Martín MJ (2009) Key website factors in e-business strategy. Int J Inf Manag 29(5):362–371
Herrero A, Martín HS (2012) Developing and testing a global model to explain the adoption of websites by users in rural tourism accommodations. Int J Hosp Manag 31(4):1178–1186
Hertel M, Menrad K (2016) Adoption of energy-efficient technologies in German SMEs of the horticultural sector: the moderating role of personal and social factors. Energy Effic 9:791–806
Ho J (1997) Evaluating the World Wide Web: a global study of commercial sites. J Comput Mediat Commun. doi:10.1111/j.1083-6101.1997.tb00066.x
Hong W, Zhu K (2006) Migrating to internet-based e-commerce: factors affecting ecommerce adoption and migration at the firm level. Inf Manag 43(2):204–221
Hohmeier KC, Borja-Hart N (2016) Are patient's simply unaware? A cross-sectional study of website promotion of community pharmacy adherence-related services in Tennessee. Res Social Adm Pharm 12:1004–1009
Hosseini MR, Banihashemi S, Chileshe N et al (2016) BIM adoption within Australian small and medium-sized enterprises (SMEs): an innovation diffusion model. Constr Econ Build 16:71–86
Javalgi RRG, Todd PR, Johnston WJ et al (2012) Entrepreneurship, muddling through, and Indian internet-enabled SMEs. J Bus Res 65:740–744
Kang YS, Kim YJ (2006) Do visitors interest level and perceived quantity of web page content matter in shaping the attitude toward a website? Decis Support Syst 42:1187–1202
Karayanni DA, Baltas GA (2003) Web site characteristics and business performance: some evidence from international business-to-business organizations. Mark Intell Planning 21:105–114
Kim H, Niehm LS (2009) The impact of website quality on information quality, value, and loyalty intentions in apparel retailing. J Interact Mark 23:221–233
Kim S, Stoel L (2004) Dimensional hierarchy of retail website quality. Inf Manag 41:619–633
Kim H, Fiore AM, Niehm LS, Jeong M (2009) Psychographic characteristics affecting behavioral intentions towards pop-up retail. Int J Retail D Manage 38:133–154
Kim C, Park S, Kwon K et al (2012) How to select search keywords for online advertising depending on consumer involvement: An empirical investigation. Expert Syst Appl 39:594610
Kittilaksanawong W, Ren Z (2013) Innovation capability building through intermediary organizations: cases of manufacturing small- and medium-sized enterprises from China's Zhejiang province. Asian J Technol Innov 21:62–79
Konings J, Roodhooft F (2002) The effect of E-business on corporate performance: firm level evidence for Belgium. De Econ 150(5):569–581
Laffey D (2007) Paid search: the innovation that changed the web. Bus Horiz 50:211–218
Lamontagne ND (2015) Academic-industry partnerships: preparing tomorrow's plastics professionals. Plast Eng 71:14–17
Lee Y, Kozar KA (2006) Investigating the effect of website quality on e-business success: an analytic hierarchy process (AHP) approach. Decis Support Syst 42:1383–1401
Li F, Du TC (2017) Maximizing micro-blog influence in online promotion. Expert Syst Appl 70:52–66
Lian T, Yu C, Wang Z et al (2016) The evaluation study on tourism websites: from the perspective of triangular intuitionistic fuzzy multiple attribute group decision making. J Appl Stat. doi:10.1080/02664763.2016.1266466
Lombard M, Snyder-Duch J (2001) Interactive advertising and presence: a framework. J Interact Mark 1:56–65

Lopez-Nicolas C, Soto-Acosta P (2010) Analyzing ICT adoption and use effects on knowledge creation: an empirical investigation in SMEs. Int J Inf Manage 30:521–528
Luo W, Cook D, Karson EJ (2011) Search advertising placement strategy: exploring the efficacy of the conventional wisdom. Inf Manag 48:404–411
Matovic B, Zivic F, Mitrovic S et al (2016) Ultra-high pressure densification and properties of nanostructured SiC. Mater Lett 164:68–71
McCole P, Ramsey E (2005) A profile of adopters and non-adopters of eCommerce in SME professional service firms. Australas Mark J 13:36–48
Mehrtens J, Cragg PB, Mills AM (2001) A model of Internet adoption by SMEs. Inf Manag 39:165–176
Mitrovic S, Adamovic D, Zivic F et al (2014) Friction and wear behavior of shot peened surfaces of 36CrNiMo4 and 36NiCrMo16 alloyed steels under dry and lubricated contact conditions. Appl Surf Sci 290:223–232
Ngai EWT (2003) Selection of websites for online advertising using AHP. Inf Manag 40:233–242
Ordanini A, Parasuraman A, Rubera G (2014) When the recipe is more important than the ingredients: a qualitative comparative analysis (QCA) of service innovation configurations. J Serv Res 17(2):134–149
Oxley M, Miller J (2000) Capturing the consumer: ensuring website stickiness. ADMAP 35(7): 21–24
Pakkala H, Presserb K, Christensenc T (2012) Using google analytics to measure visitor statistics: the case of food composition websites. Int J Inf Manag 32:504–512
Ramanathan R, Ramanathan U, Hsiao HL (2012) The impact of e-commerce on Taiwanese SMEs: marketing and operations effects. Int J Prod Econ 140:934–943
Rayport JF, Jaworski BJ (2004) Introduction to e-commerce. McGraw Hill/Irwin, Boston
Redoli J, Mompó R, García-Díez J et al (2008) A model for the assessment and development of internet-based information and communication services in small and medium enterprises. Technovation 28:424–435
Rizos V, Behrens A, van der Gaast W et al (2016) Implementation of circular economy business models by small and medium-sized enterprises (SMEs): barriers and enablers. Sustainability 8 (11):1212. doi:10.3390/su8111212
Roselius T (1971) Consumer rankings of risk reduction methods. J Mark 35(1):56–61
Rosen DE, Purinton E (2004) Website design: viewing the web as a cognitive landscape. J Bus Res 57:787–794
Scaglione M, Schegg R, Murphy J (2009) Website adoption and sales performance in Valais' hospitality industry. Technovation 29(9):625–631
Scarpellini S, Aranda A, Aranda J et al (2012) R&D and eco-innovation: opportunities for closer collaboration between universities and companies through technology centers. Clean Technol Environ 14:1047–1058
Soto-Acosta P, Popa S, Palacios-Marqués D (2016) E-business, organizational innovation and firm performance in manufacturing SMEs: an empirical study in Spain. Technol Econ Dev Eco 22:885–904
Soto-Acosta P, Colomo-Palacios R, Popa S (2014) Web knowledge sharing and its effect on innovation: an empirical investigation in SMEs. Knowl Manag Res Pract 12:103–113
Statistical Office of the Republic of Serbia (2011). Use of ICT technologies in Serbia. http://webrzs.stat.gov.rs/WebSite/repository/documents/00/00/43/62/PressICT2011.pdf. Accessed 17 July 2014
Syrkiewicz-Switala M, Romaniuk P, Ptak E (2016) Perspectives for the use of social media in e-pharmamarketing. Front Pharmacol 7:445
Tajonar A (2014) How to start a biotech company. Mol Biol Cell 25:3280–3283
Teo TSH, Pian Y (2004) A model for Web adoption. Inf Manag 41:457–468
Tiago MT, Tiago F (2012) Revisiting the impact of integrated internet marketing on firms' online performance: European evidences. Procedia Technol 5:418–426
Triandini E, Djunaidy A, Siahaan D (2017) A maturity model for e-commerce adoption by small and medium enterprises in Indonesia. J Electron Commer Organ 15:44–58

Vencl A, Rajkovic V, Zivic F et al (2013) The effect of processing techniques on microstructural and tribological properties of copper-based alloys. Appl Surf Sci 280:646–654
Vila N, Kuster I (2011) Consumer feelings and behaviors towards well designed websites. Inf Manag 48:166–177
Wang Y, Ahmed PK (2009) The moderating effect of the business strategic orientation on eCommerce adoption: evidence from UK family run SMEs. J Strateg Inf Syst 18:16–30
Webb HW, Webb LA (2004) SiteQual: an integrated measure of website quality. J Enterp Inf Manag 17:430–440
Weinberg BD, Pehlivan E (2011) Social spending: managing the social media mix. Bus Horiz 54:275–282
Weltevreden JWJ, Boschma RA (2008) Internet strategies and performance of Dutch retailers. J Retail Consum Serv 15(3):163–178
Whitehead J (2008) IPA backs ads in face of downturn. Marketing 9:4
Woodside AG (2015) Constructing business-to-business marketing models that overcome the limitations in variable-based and case-based research paradigms. J Bus-to-Bus Mark 22(1–2):95–110. doi:10.1080/1051712X.2015.1021589
Woodside AG (2014) Embrace perform model: complexity theory, contrarian case analysis, and multiple realities. J Bus Res 67:2495–2503
Woodside AG (2013) Moving beyond multiple regression analysis to algorithms: calling for adoption of a paradigm shift from symmetric to asymmetric thinking in data analysis and crafting theory. J Bus Res 66:463–472
Wu PL, Yeh SS, Huan TC et al (2014) Applying complexity theory to deepen service dominant logic: configural analysis of customer experience-and-outcome assessments of professional services for personal transformations. J Bus Res 67:1647–1670
Zheng J, Caldwell N, Harland C et al (2004) Small firms and e-business: cautiousness, contingency and cost-benefit. J Purch Supply Manag 10:27–39
Zivic F, Babic M, Grujovic N et al (2013) Influence of loose PMMA bone cement particles on the corrosion assisted wear of the orthopaedic AISI 316LVM stainless steel during reciprocating sliding. Wear 300:65–77

Part II
Case Studies

The Profile of Researchers Moving Towards Scientific Entrepreneurship

Xavier Marti, Ignasi Fina, Gustau Catalan and Andreu Veà

Abstract We review some of the skills that scientists who have become scientific entrepreneurs share. We advocate that it is possible to include technology transfer as a part of the scientific investigation not only as a collateral funding source but also as a way to reinforce the scientific advances and resulting communication. Traditionally, the technology-transfer process of scientific results has been mediated mainly by patents, even before the first prototype testing has been performed. Instead, we discuss the critical benefits of fast prototyping and develop strategies how to adequately monetize and establish the price of the products even at the prototype level. We also list several common failures of a technology-transfer project and some preventive actions to avoid them.

Keywords Technology transfer · Entrepreneur

X. Marti (✉)
IGS Research, C/La Coma, Nave 8, 43140 La Pobla de Mafumet (Tarragona), Spain
e-mail: xavi.marti@igsresearch.com

I. Fina · G. Catalan
Catalan Institute of Nanoscience and Nanotechnology (ICN2), CSIC and the Barcelona Institute of Science and Technology, Campus UAB, 08193 Bellaterra, Barcelona, Spain

I. Fina
Institut de Ciència de Materials de Barcelona (CSIC), Campus de Bellaterra, 08193 Bellaterra, Spain

A. Veà
Board at the Internet Society (ISOC-Es), Santiago de Compostela, Spain

A. Veà
Digital Champion for Spain, European Commission, Barcelona, Spain

J. Peña Dopazo and F. Zivic (eds.), *Supporting University Ventures in Nanotechnology, Biomaterials and Magnetic Sensing Applications*,
DOI 10.1007/978-3-319-61237-9_7

1 Introduction

While one can find extensive support for setting up and managing many types of spin-offs and start-ups, the actual specific process for creating a small legal entity comprising one or two academics, plus one or two non-academics (the typical starting size of a business attempt based on a scientific discovery) is, at the moment of writing, not well described. And yet, reality points to the fact that the profile of a successful academic who interested in technology transfer calls for playing a critical role in raising funds and thus stimulating novel research and progress. The underpinning idea is simple: academia is the source *par excellence* for creation and discovery. Therefore, one should expect it to be a good seedbed for innovation. In practice, though, there are many issues that need to be considered. In the next sections, we will review a few of the factors that either limit or empower the transformation of academic creations into commercial products. **The ultimate goal is to unravel whether a scientific group can continue to produce high-quality science while covering part of its operational costs by *participating* in (not necessarily *leading*) the technology transfer chain**.

This chapter is organised as follows: in Sect. 2.2 we discuss some fundamental characteristics of researchers' profile; in Sect. 2.3, we remark about the critical role of developing a prototype in the early stages of the technology-transfer process; in Sect. 2.4, we address the process of pricing and how it is perceived by the researcher—who usually underestimates the value (and price!) of time and knowledge; in the last chapter, we enumerate a few actions that can create the appropriate environment in your research group to initiate a successful spin-off company.

2 Scientific Skills in an Entrepreneurial Environment

2.1 Creativity: Mind the Gap

The first factor to consider stems from the word "creativity". Scientists and researchers are by default creative. They are paid to come up with new ideas, and the current parametrization of performance in academia rewards primarily "novelty and impact". If a scientist "A" has worked for many years in the study of a phenomena of interest and reaches very sound conclusions, that person would be ready to submit the work to a high-profile (high-reward) journal. However, if, in the meantime, somebody else "B" publishes a paper with a similar idea but containing much less detail elsewhere, the mere existence of that second, perhaps sloppier work, seriously threatens the publication (and the reward) of A's work. The work of B has taken away the novelty aspect, which is the main asset. As a result, the scientific community may be unconsciously disregarding some areas of research just because somebody else has also covered it. In other words, this scenario

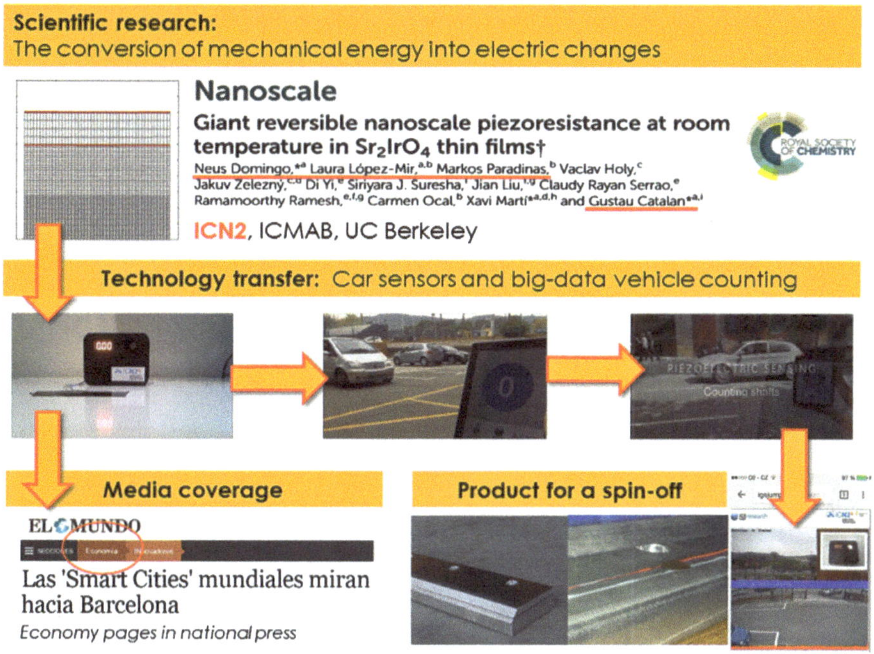

Fig. 1 Sketch of the initial steps for setting-up an academic spin-off. It is summarized as follows: paper, prototype (the seminal ingredient for technology transfer), strong dissemination and the first sale. The case corresponds to the process initiated after the publication of a scientific article (Domingo et al. 2015) and the initiative by the Institut Catala de Nanociencia i Nanotecnologia to develop a prototype product related to the activity at the research institution

favours the initiation of topics rather than consolidation of pre-existing ones: "first" is better than "best". This algorithm has two consequences: first, it favours the reduction of the work force devoted to any specific topic and, second, it reduces the lifetime of a topic to a timescale that might prevent the technology-transfer staff to realize the economic potential of the scientific discovery (Fig. 1).

Let us illustrate this situation with an example. Imagine there is a potential novel application that relies on two published phenomena already covered in scientific papers. A third scientist, proficient in at least one of those two areas of expertise, detects this opportunity and is capable of writing a third manuscript on the "smart combination" of the former ideas. Due to the actual gap between science and technology-transfer languages, the profile of the third scientist is perhaps the only one capable of bridging the scientific research and technology transfer. However, whenever he/she inspects the performance metrics, it turns out that this type of manuscript cannot be submitted to the group of journals he/she is familiar with. Indeed, it is very hard to know what to do with this type of literature that demands expert scientific input but delivers technology transfer. Interestingly enough, many scientists admit that they enjoy exploring beyond the boundaries of the canonical domains of academic science. There is a genuine attraction to know what could happen beyond the papers. However, at the moment of writing, *beyond the papers work* is not sufficiently parametrized in order to become an item of the performance examination.

Here, it is important to emphasize that *only* filing patents (currently the default parameter to judge the technology-transfer record of a scientist) is not the optimal solution. Patents are a quantitative way to assess the ability to transform scientific ideas into documents that are of interest in the subsequent stages of the technology transfer. But, again, filing patents is not enough to finally transform something that lies in the column of "costs" into something that has potential to shift towards the column of "income". The ultimate purpose of the investment in having a technology-transfer department is this shift between columns. Otherwise, we help reducing the gap but, there is still no connection between the creativity and the monetization. As we outlined in the example, there must be an active scientific participation in the technology-transfer process and that should be actively credited, parametrized and rewarded—even in the event of failure. In a similar approach as used for quantifying the hours devoted to teaching activities, the time spent in attempting technology transfer may be valued in terms of *credits*. Of course, quality control should apply here to avoid the extreme cases in which some individuals hack the system; just spending time on something is of course not an added value per se, but what we mean is that the quality of the technology-transfer actions has to be audited independently as to whether they ultimately achieve the goal of making money. Only then would a substantial number of scientists care at all about the technology transfer, and, by the law of probabilities, more viable businesses would emerge. Observing from the perspective of an entire research institute, one realizes that, by increasing the amount of technology-transfer processes, then there is a higher probability of success of some of them. Subsequently, having one success story helps empowering the emerging surrounding projects, plus it provides valuable in-house knowledge about how it happened and what local factors did or did not help.

The two take-home messages from this section are that (i) From a technology transfer point of view, one should be careful not to reward novelty at the expense of expertise: real expertise is the most solid foundation for a viable business, (ii) For scientists, there is currently a large gap in professional incentives between discovering new things and attempting to monetize the discoveries. Until that gap in professional rewards is addressed, entrepreneurial scientists will be as rare as hen's teeth (Fig. 2).

2.2 Flexibility: Bend or Break

The apparent rigidity of scientific logic often clashes with the need for a quickly adapting rhythm for an emerging start-up and, at the very end, the actual needs of your first customer.

Fig. 2 Summary of the skills of a scientific entrepreneur

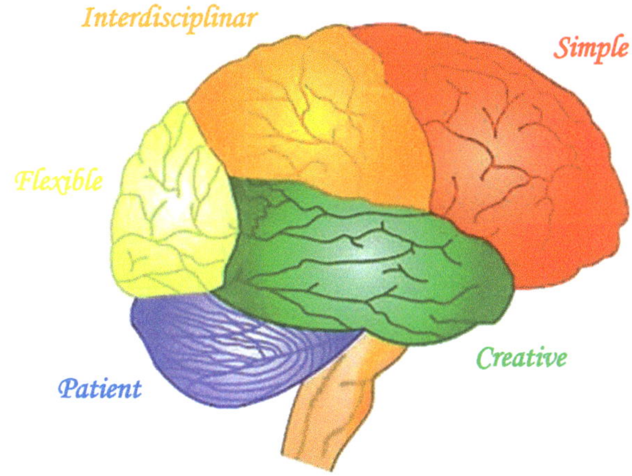

One of the common barriers to overcome is the weight of the labels that stick to one's work. The ideal case is, of course, that in which a scientist makes a significant discovery in one area and his/her very same samples are embedded in million-dollar market applications the next day. But, as we learned from thermodynamics, ideal laws apply to some gases and in a very few more selective cases.

On these grounds, it is important to be open-minded when being involved in a technology- transfer process carried out on a research-institute scale. It may happen that, even if your latest discovery is labelled as "nano" and has produced a notable impact on the scientific community, it inspires a first product which is "micro" or even "macro". So what? If it was inspired by your research and is going to make money, then that is good technology transfer. It may even happen that the business will come full-circle and deliver a product in the future.

Having the ability to focus and defocus, to appreciate the gray-scale levels, and, having the capacity to react flexibly to these subtle and changing needs is a fundamental attribute towards identifying and following a fruitful path. One common mistake is the impulse to wait for a too-perfect first prototype. As we shall see in subsequent chapters, the first prototype has to come out as soon as possible and, most likely, it will be a non-working prototype. Accepting that a non-working prototype is better than no prototype is the first "stress test" of the flexibility of the participants in the business venture, and one that is not easily passed by any self-respecting rigorous scientist. But the alternative, as we shall see, is worse: no business at all.

Please let us also emphasise an important point that may otherwise be misconstrued from we have already written. Honesty is the prize asset of any business, and, once trust is lost, it is never recovered. Thus, a non-working prototype may be acceptable, necessary even, as long as everybody in the chain, including the customer, is fully aware of the nature of said prototype, i.e. it is a prototype built precisely to calibrate what are the things that do not work and need to be fixed. NEVER try to pass a fake result or a non-working device as a working one. Being flexible does not mean being bent.

2.3 Interdisciplinarity: Easier Said Than Done

In the previous section, we presented the example of a scientist connecting two already existing ideas. Usually these two ideas are coming from different disciplines. Once more, many scientists feel attracted to walk back and forth adjacent fields of research but, at the moment of putting in written form the outcome of these interactions, it is hard to locate the right forum to submit the manuscript or, harder still, the right panel to which submit the application for funding. It is also difficult to trace a cooperation between institutions sitting in opposite locations of the knowledge map.

Encouragingly, in the past years, some promising fields have been opening avenues between various different disciplines: for instance, biology, medicine, physics, pharmaceutical etc. In some of these cases, the strong budget provided by some industrial partners helps softening the eventual difficulties. However, many emerging proposals for connections still do not receive this initial attention, thus hindering growth. In this aspect, besides the push by private monetary inputs, we have observed in our experiences that most evaluation panels are scrutinizing only one shore of an interdisciplinary proposal, and the added benefit of cross-field bridging itself is not seen as a bonus.

For empowering the entrepreneurial character, an important pedagogical task must be performed to turn the evaluation panel's sensitivity towards the value of interdisciplinary projects. On the plus side, combining two ideas that are already thoroughly validated in two distinct fields constitutes a low-risk enterprise that may better attract private investors precisely because it is temptingly simple—and brings us to the next section.

2.4 Simplicity: Less Is More

There is an actual risk of wanting to validate the novelty of too many ingredients when thinking about creating academic spin-offs. Yet, outside the circle of academic research—where novelty is a requirement, as discussed in previous sections—successful start-ups and businesses are, in general, based on the smart and innovative assembly of pre-existing building blocks (such as integrated circuits, sensors, communication devices, third parties services etc.) that are robust, well-documented and, usually, complemented by extremely rich user communities. Besides, there is a virtually unlimited supply of any of such objects, that is, the variety of already available "atoms" is not an issue. In other words, it is not so important to come up with new shapes for the "lego" blocks, but with new and useful ways to put them together.

In our experience with technology transfer, the parts/product availability and the algorithm universality—the reproducibility of the output in all conceivable, and sometimes inconceivable, usage circumstances—have always been among the key limiting factors. Therefore, the "keep it simple" becomes a fundamental point for the survival of academic spin-offs. No matter how big the scientific discovery and its impact factor, the path to reaching a satisfied customer is long and circuitous, and "traveling light" in terms of simplicity and availability of the building blocks becomes a huge bonus. Be also aware that a compromise between scientific novelty and technological simplicity is necessary in order to be successful.

This is particularly true for the first prototype, which needs to be released as soon as possible! Remember: a spin-off is not an extension of the laboratory. It does not reward the same virtues or with the same bonuses. It does not operate according to the scientific method but according to business logic; sometimes it is even contrary: a good scientist will try to invent as many new things as possible, but in a real business the fewer pieces you need to invent or make yourself, the simpler it becomes to make the prototype or the product.

We want to emphasise once more the ultimate purpose of the investment in technology transfer: to feed the column of "income" of the research institution spreadsheets while minimising the impact in the scientific activity. In our times of scarce resources, no funding strategy must be a priori disregarded. In this regard, we admit that simplicity and low-risk clash with novelty and impact in the malicious comparison between the wishes of capital and science. However, **our vision is that, if properly irrigated, smart and well-aimed technology-transfer activities at the scale of a research institution will generate income that will sufficiently compensate for the time dedicated to it.**

Time, by the way, is a scarce resource for a research scientist and, ironically, it is increasingly occupied by administrative tasks. There is a need for mutual trust and patience to reach a successful point, as we discuss in the next section.

2.5 *Patience is the Mother of Science*

Academic spin-offs may exhibit slower development speeds, and one of the causes will be rarely fixed as it stems from the very heart of the constituting elements: the scientists. A message to the non-academic co-owner or technology-transfer staff: be patient and realistic. These moderate-speed scales are caused partly by lack of sufficient "keep it simple" efforts (see previous section), but also due to the fact that the co-owners are generally willing to keep their academic appointments, which requires them to put most of their efforts into research outputs (scientific papers) rather than business ventures with uncertain futures. This is natural: scientists know very well how science works and have reasonable security that their time invested in the lab will be rewarded by an article within a few months' time; it makes sense to keep doing what one is good at and satisfies one's ego as well as one's employers.

Although canonical entrepreneurs' manuals dictate that setting up a start-up cannot be a hobby or a part-time job, the reality is that in academic environments this would be an exception—despite some celebrated cases on the media—that someone risks a permanent position, based on scientific output, by putting all his/her time into a business venture. Furthermore, we cannot conceive a society model based on a continuous decanting of science-makers into business-makers. Moreover, we would set big and bright warning signs at the poor intersection between the native skills set of a scientist and a chief executive officer. **We strongly believe that the model to turn the technology transfer into income at a research institution scale must <u>not</u> be based on demanding that scientists become CEOs**.

There are many corollaries to the previous statement. Let us sketch one of them that is important also for the next section. At the very early stages of the academic spin-offs, the development of the first prototype has to be realised by specialised and full-time dedicated staff, either hired within the group, if such position exists, or outsourced to a third company. Prototypes are needed for the idea's validation, but also are essential for demonstration purposes: as they say, "seeing is believing", and the publicity value and attractiveness to business investors of a tangible prototype, as opposed to a sketch on a piece of paper, cannot be overstated. Once that prototype is made, visible and tangible, one can immediately collect the first potential customers' or investors' feedback. Therefore, we emphasise that **it should not be the scientist (despite her/his passion and skills for fabricating experimental set-ups) who actively fabricates the first prototype**. Instead, the prototype must be designed by someone whose work is entirely parameterised by the delivery of the tangible object. Our golden rule reads: scientists must do science and only support the adjacent steps. In this chain, everybody is expected to keep his/her seat and use the arms to touch the adjacent participants.

Who pays? We see every day that academic institutions invest heavily in the hiring of new staff, sponsoring visits by high-level scientists, acquiring expensive high-tech research equipment, outsourcing complex lab construction works, organizing workshops and macro-conferences etc. In this context, why should supporting the early stages of a spin-off creation be an issue? If one reads what some academic institutions claim on their webpages, it seems coherent that they would pour some money into a prototype-making department. In the next section, we discuss the relevance of having an early prototype.

3 Prototype: Demonstrate or Die

During our careers as scientists, we have heard many times the following advice: "publish or perish". This means that, if you do not produce articles at the end of your research, it is as if you had never done the research, and your academic career is of no value to the community. The pressure to publish becomes even more intense when one acquires that most scary of academic positions, that of "senior postdoc", where "publish or perish" is combined with a new demand: now is the time to "make or break"; you either make something new and important and get that permanent appointment, or you might as well change career. Unfair though it may seem to the academic staff involved in this "publish or perish" rat race, the lessons can be directly applied to the start-up environment, where failure to make a demonstration device means failure to attract any interest. Here we propose that the golden rule for the budding academic entrepreneur should be "demonstrate or die".

Ironically enough, many of the wannabe start-ups we have come across also fail to achieve the point of "make or break". This is because most fail at even the most basic step of "making", i.e. producing a proof-of-concept prototype. There are zillions of ideas, but how many of them have been validated by testing an affordable, viable-looking prototype? Here, there may be an element of fear in knowing the truth that has to be addressed. In some circles, it is preferred to never know in which box hides Schrodinger's cat. It is preferable to hope that it may be alive, or at least, quantum-mechanically half-alive (whatever that means), than to open it and find out that it is dead. However, in order to make the technology transfer profitable, one has to inspect both boxes as soon as possible—and be ready to leave the room and get a new pet.

We cannot emphasize enough the importance of developing a prototype. Even if in the first iterations it is a non-working prototype (let's call it a NWP), it will be a coherent collection of materials, circuitry and software that will help all the surrounding actors to stay with the motivational high tide. They say that seeing is believing, but in fact it is even truer that touching is believing. Perceiving progress in your co-workers via a work-in-progress prototype is far more encouraging and exciting than the eerie silence of a non-existing prototype that never makes it off the drawing board. In order to stimulate the creation of prototypes, the research institution can use its technical workshops, engineering staff and even strong marketing department to boost the visibility of the academic paper that was the seed of the current NWP. This is a crucial step that helps realising at an early stages the major flaws of the envisaged application and, on a different page, it strongly supports future grant applications.

An early fail saves you time and teaches you valuable lessons for the next attempt. A "smart failure" is one in which you have been able to parametrise the events that led you to fail. That is, in fact, at the very core of the scientific method! Failure to fail is a failure in itself, as it means you did not even try. Therefore, move quickly to the stage in which you can develop a NWP. Get partners and full-time manpower for that too: don't pretend that a single person can cover everything, from the creative scientific discovery up to the fine selection of nuts and bolts, for the prototype.

At this point, we would like to quote that: "Never tried. Never failed. No matter. Try again. Fail again. Fail better".

4 Money: The Price of Everything and the Value of Nothing

We scientists are used to work for fun, or for free, even when it is not fun: for example, we peer-review other people's articles without charging a cent, even though it is a thankless, time—consuming and sometimes even emotionally charged task. Why we do this remains a mystery that could be the subject of a separate chapter in a psychology book. Moreover, we accept without much protest that the value of our work is decided by a complex and somewhat stochastic process, whereby we submit our work to anonymous editors and referees who may be either friends or foes, colleagues or competitors, who will decide its fate. When the article is published (it always gets published, of course; the question here is not "if" but "where"), the price tag of the article is given by the so-called "impact factor" of the journal, which is a number that is not determined by the quality of the article itself, but by the number of citations to other people's articles previously published in the same journal. And, as well as this uncontrollable impact factor, there is of course the subtle self-regulated equilibrium of scientific bit-coins related to co-authorships, co-operations, co-participation in projects etc.

So far, we have mentioned words such as work, price, value etc. When setting up an academic spin-off, the meaning we give to such words may need a radical change. And, definitely, they must be variables under more direct control than the impact factor of an academic paper.

The academic life relies on costs (salaries, etc.) being paid from projects from which one invests a small portion of time and resources to generate the seeds of the next funded projects. But, writing projects, then writing progress reports on the projects, and writing also refereeing reports on other peoples' projects takes time. A lot of time. And this is of course in addition to the time that should be spent in the lab, or advising students, or writing scientific articles or revising other people's scientific articles. In this circle, one observes that the demands on a scientist's time for "community obligations" stretches more than in a sci-fi movie—with the net result of having to work long hours on paperwork, many of those over the weekend or during holidays. Unless, of course, the scientist decides to take his holidays seriously, in which case what suffers is the actual research time. And yet, one rarely puts (or knows) the public price of one's own work or research, because it all flows in a natural equilibrium: we sit on scientific panels and editorial boards, assess grant proposals, review articles and lend our time, knowledge and equipment for free because, and we expect that other scientists will do the same. And, mostly, they do! Professional science is as close to the communist utopia as one can get in this world.

On these grounds, the external perception turns out to be that we, scientists, can give away substantial portions of their time without too much hesitation. While one could put to work one senior and two post-docs with one good idea, such magic would never work with a car mechanic. Time is money—do not forget this.

Now we move towards the opposite end. A spin-off requires a monetary investment by your home institution and/or investors and cannot accept time, papers or invited talks as currency; the banks will not accept that as a return on their loan. Therefore, one has to put a price sticker on one's work, be it a salary for oneself, the value of a patent license or spin-off company or the actual final sales price of a product made by the spin-off. Unless you are in very specific business model or your spin-off is primarily part of a joint research program with external companies who take the lead, your customer needs to know *how much is it?*

So, how much is it? What is the aggregate cost of your time, expertise, equipment, ideas, raw materials, student-time etc.? What will be the price tag on the product, spin-off company, consulting fees or license that you are selling to investors or buyers?

The first instinct says something simple: sum up all expenses, add a margin of more than 20% and deliver this number. This 20% is just a very general estimate, and it may depend on each particular field. Let's call this price P1. Computing P1 seems relatively easy because all information needed is at your hand and needs no information about the customer (suspicious… isn't it?). Even then, a proper calculation of P1 is by no means easy, and, if it is done properly, one will quickly realise how we scientists tend to underestimate our costs; P1 is invariably much larger than our first guess. But, even if we calculate P1 correctly, this is still not the optimal way; ignoring the customer in the equation is a BIG mistake. Inspired by discussions with Dr. Carlos Frontera during the technology-transfer sessions at Institut Català de Nanociència i Nanotecnologia, we learned that it is crucial to analyse what your customer wants, how your customer uses your product or services and estimate *how much money is your customer saving* by using your product or services. Divide that last amount by a factor of two and you can have one good estimate of how much your customer would be happy to pay without too much negotiation. We will call this amount P2. We see here that there is an asymmetric situation: we know first-hand how much our product costs to make but we would like to know how much the customer is happy to pay. Meanwhile, the customer is in the opposite situation: they know very well how much they are willing to pay, but only have a faintest guess of how much it cost us to make the product. They could of course calculate our production costs themselves, but that is time-consuming and, outside of academia, time is money, so they are quite happy to ignore our costs and pay "over the odds" if they are still saving a lot of money.

So, it falls again to us, the entrepreneurial scientists, to invest time and calculate our potential buyer's purchase potential, what we call P2.

Calculating P2 is in reality a bit complicated. For a start, looking at the price of the competition is important. However, there may not be competition at all (that is the case for truly ground-breaking technologies, though these are rare). Also, besides the direct savings incurred by buying your product, one should also consider eventual additional earnings produced by your product. Also, when looking at protracted time scales (a decade is a good order of magnitude), what will matter for the customer is the so-called ROE (return on equity). If the acquisition of your product renders an ROE 20–25%, it is a great deal for your customer, while the minimum level would be 10%. If you do your homework well, by the end of the process, you will not only know what your minimum price is, but also what is your customer is ready to pay, i.e. you know both P1 and P2 with reasonable accuracy.

As your instinct says, to operate your spin-off safely, P2 should be larger than P1 in the long term. In fact, you will often find out that P2 is not just larger than P1, but MUCH larger, i.e. P2 $\gg$ P1. The reasons for such inequality are many: first, as outlined earlier, scientists tend to underestimate how much is the hourly rate of a highly qualified and experienced professional, but businesses know this figure very well, because that is the rate that they are currently paying and/or earning; second, infrastructure, travel costs and accrued knowledge are also typically underestimated: making a professional move from one place to another multiplies the hourly rate; using that Excel datasheet that took years of aggregate refinements by smart PhD students in order to quickly calculate one parameter at the push of a button, or having all your lab sorted out ready for giving an answer, does not come for free—not to mention the internet connection and subscriptions used. Sure, it is all there ready to use at the drop of a hat and is cheap to use *now*, but the time it took to build it was substantially longer than that proverbial drop of a hat, and way costlier. As the saying goes: when the plumber comes to fix your boiler and takes 5 min to push a button, but charges the equivalent of a day's hard labour on a farm, he or she is not charging for pushing that button, but for knowing which button to push. Knowledge, literally, is money.

An informal survey carried out by Xavi Marti in the last scientific conference he attended rendered a very interesting result. While chatting with both scientists and exhibitors, the latter repeated many times words along the lines of "complexity" and "very advanced technology which carried long development times by highly qualified personnel", and of course "patented" so that one could not get any more details. On the contrary, scientists described their findings as "simple and very cheap", carried out "in a very short-time scale", and in some cases explicitly mentioning "the hard part was developed by a master student". This is not a judgment or a confrontation but just a description. **And the bottom line is: when preparing your spin-off spreadsheets, be ready to charge high hourly rates**.

We scientists are used to being happy with payment in species, so we usually underestimate the costs of our working hours, and businessmen are sometimes derided for knowing the price of everything and the value of nothing. This is of course a stereotype, but scientists fall almost into the opposite stereotype: they know the value of everything, but the price of nothing. This attitude towards money must be changed. We scientists sit at the cusp of human creativity, and we have a high value because we are the door to the new building blocks. It is, therefore, our responsibility to actively participate in the technology-transfer chain to turn these ideas into tangible progress and, likewise, it is our right to be fairly paid for it.

5 Ingredients for Success

Unfortunately, there is no such unique recipe for success concerning the creation of academic start-ups. Here, we enumerate a few actions that have been observed in successful cases and might carry some degree of universality. The ideas shown here apply, primarily, to projects growing from absolute zero. Of course, if substantial pre-existing elements are available, some of those actions are not needed, and the entire perspective may be slightly different.

First, it is as important to identify one area in which it is possible **to generate, relatively quickly, a first invoice** of about 10,000 euro selling or renting the first prototype (or service). The reason for this is because it is important to show frequent incremental steps in environments without a protracted record of entrepreneurship. That helps the entrepreneurs and shields them from discouraging inputs from sceptical nearby observers. Though such personal feelings seem irrelevant in many circles, the scientist's ego is often fragile and ignoring the *nearest neighbour interactions* can lead to a bad ending. In this initial business choice, as we mentioned in Sect. 2.3, be ready to be flexible: there is the need to take off, and once you are at cruising altitude you can fine tune your heading. But if your flight does not take off quickly, there will be nothing to assume the flight path. Finally, having completed the circle of cashing your first invoice (and learning how the legal, bureaucratic and tax issues work in practice), you will be better suited for repeating this procedure and, specially, having a clearer picture of how to price your stuff.

Second, **identify and nominate a person clearly responsible for manufacturing the first prototype**. Manage to hire a person who will carry the flag and will be the contact person. There is nothing worse than a prototype project without a visible responsible head and a deadline. One cannot rely on the agenda fluctuations of researchers, with teaching and traveling tasks popping up unexpectedly. That would delay the official release of the prototype and cool down the interest of the targeted first customer. Another important aspect is that the prototype has to be able to demonstrate a principal functionality but, at this stage, by no means is expected to be perfect. Beware of the passion for putting off the release because of the goal of infinite perfection: do not let "best" be enemy of "good".

Third, **take the marketing tasks seriously**. Most probably, no customer will be aware of your h-index (or actually even know what it means), but you can use your scientific publications and your institutions marketing department to boost the echoes of your latest discovery. Be ready to post frequently on social networks and attract the attention of local media to raise the relevance of your work. Notice that, as a valuable by-product, you are working to help your scientific work to be disseminated.

Fourth, **validate your ideas as soon as possible**. It is important to identify a failure because that will save you a lot of time. Find as many as possible testers of your prototype. Insist to get paid for renting this prototype. Dealing with completely free stuff does not trigger the necessary interest from both sides to truly validate the process. Within the validation process, one has to identify a small or medium enterprise that is capable of manufacturing the first hundred units. This is a very valuable step because the first big issues on the scaling up will be identified here at very early stages. On the other hand, this activity puts you into contact with actors who are not very sensitive to the scientific aspects and rather will inspect the practicalities from an unbiased point of view.

After these initial actions have been completed, one is ready to face the challenge of a serious round of fund raising from private investors. Without having a tangible validated prototype, it is much harder to explain the potential impact of your ideas. As a bonus, you will have contributed strongly to the dissemination of your scientific work, the public image of your research institution and you will have learned the order of magnitude of work involved in a co-operation with technology transfer—and most probably you will have continued to published papers and attend conferences.

Finally, we want to emphasise one important, and often forgotten, benefit of participating in the technology-transfer chain. To boost any field of scientific research, one doubtlessly needs powerful and eye-catching realistic introductions to either manuscripts or grant applications. If one branch of research is particularly original, then it is very likely that no large enterprises or commercial products can serve as a major marketing claim for funding your research. In this situation, don't give up. Go and create your own introduction by attracting your technology-transfer actors to build a simple gadget and do a great deal of marketing for you. The truth is that, if you open a new research field, nobody better than you can act as the seed actor for the technology transfer. And this activity should never hinder your scientific goals, instead it can only benefit them.

6 Summary and Conclusions

Entrepreneurship is growing, and so the entire economic system is adapting to its existence. On these grounds, academics can also benefit from some of the features traditionally ascribed to entrepreneurs. By no means does bridging between the academic and business worlds need to be achieved by a radical change of the scientific method and traditional practises. A mere change in the communication between those worlds can already benefit both sides. In this chapter, we listed a few of the differentiating skills that can help a scientist facing the challenges of dealing with entrepreneurship. In particular, we described a case where early prototyping of scientific *gedanken-experiments* is a key step to make both worlds converge, even generating money. Indeed, this step does not require large initial investments, so that it is compatible with the typical level of funding that scientists are obtaining nowadays from public sources. We conclude that, in some cases, a functional and valuable prototype development can require less money than patenting costs and, instead, generate many more benefits on the academic side.

Reference

Domingo N et al (2015) Giant reversible nanoscale piezoresistance at room temperature in Sr_2IrO_4 thin films. Nanoscale 7:3453–3459

IGSresearch: From Science to Business in the Markets of Security, Smartcity Management, and Geological Monitoring

Javier Garcés, Ignasi Fina and Xavier Marti

1 Introduction

IGSresearch Ltd. is a customized electronics manufacturing company with a strong focus in the field of magnetic sensing applications. In 2015, the company celebrated its 15th anniversary while it consolidated its transformation from, at first, a software developing company into a technology-based enterprise contracted by world-class research institutions.

During the years 2013–2015, the company settled its own large-scale manufacturing facilities, which comprise fully-automated printed-circuit-board assembly and encapsulation. On these grounds, IGSresearch Ltd. is capable of developing its own commercial products encompassing the electronic design, fabrication, and commercialisation. Alternatively, IGSresearch pursues cooperations with academic and research institutions that have just released to the public knowledge about novel functionalities. The control of the entire supply chain enables the company to sustain its operation based on its own sales, which have been eventually supplemented by grants and contracts with private investors. In recent years, the company had started exporting goods to the Eastern European and South American markets, thus consolidating a robust base for the challenging years to come (Fig. 1).

J. Garcés · I. Fina · X. Marti (✉)
IGS Research, C/La Coma, Nave 8, 43140 La Pobla de Mafumet, Tarragona, Spain
e-mail: xavi.marti@igsresearch.com

I. Fina
Institut de Ciència de Materials de Barcelona (CSIC), Campus de Bellaterra, 08193 Bellaterra, Spain

J. Peña Dopazo and F. Zivic (eds.), *Supporting University Ventures in Nanotechnology, Biomaterials and Magnetic Sensing Applications*,
DOI 10.1007/978-3-319-61237-9_8

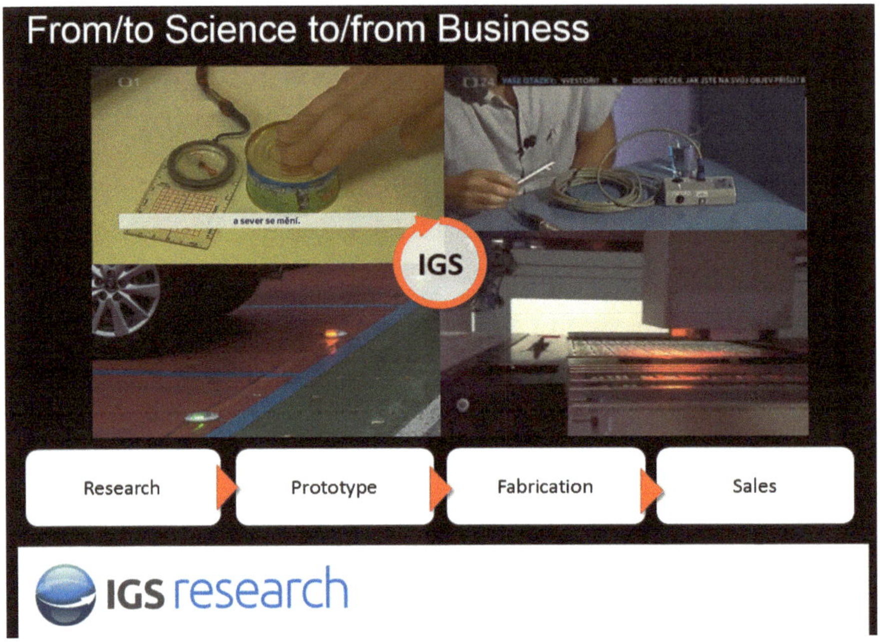

Fig. 1 IGSresearch Ltd.'s technology transfer vision: first, identify the areas of research in which exploiting their potential involves microelectronic component assembly and powerful mathematical algorithms; second, promptly develop a prototype for validation and ignite the marketing together with world-class research institutions; third, reduce dependencies on external partners in-house large-scale manufacturing facilities; fourth, generate sales and make contracts with international distributors through the know-how of a strong marketing team. Image: IGSresearch

2 Spintronics: From Hard-Drives to the Smartcity

Magnetic positioning has been facilitating automated assembly lines for the past 50 years. It is, indeed, the best cost-effective choice for an industry demanding great accuracy in the monitoring of metallic objects. Such technology was picked up by the data-storage industry in the early 1990s, and such magnetic sensors were inserted into the hard-drive read-heads. The consequences in the data storage industry were so significant that magnetic-sensing technology earned its innovators a Nobel Prize in 2007. In the past years, IGSresearch Ltd. has revisited many of the concepts learned at the nano-scale and adapted them to vehicle detection in the field of smartcities. There is one feature that makes the magnetic sensing unique in smartcity applications: the unrivalled ability to sense objects without the need for a direct line-of-sight. This is a crucial aspect involved in shielding the sensing functionality against water, vandalism, and harsh climatic environments. The international expansion of this technology has been fuelled by the installation of validation sites in world-class academic institutions, such as the University of Barcelona and The Czech Academy of Sciences (Fig. 2).

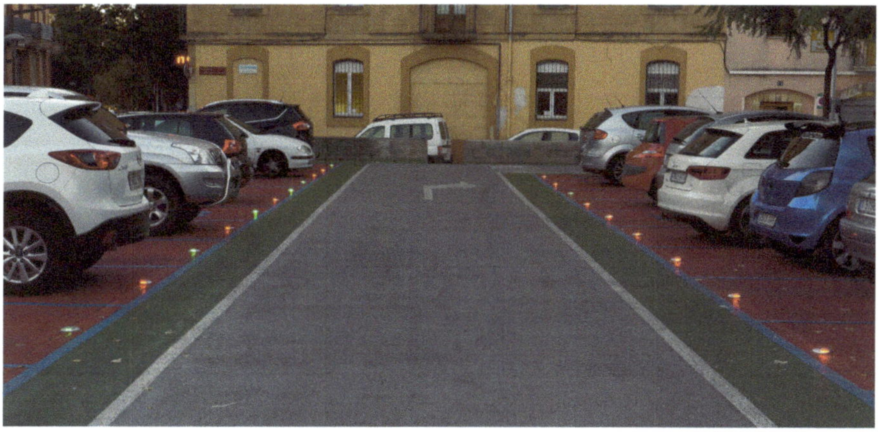

Fig. 2 IGSresearch Ltd. released the product SPINWIRE® INTEGRATED SOLUTION in 2012. Over the past few years, SPINWIRE® has become the ultimate solution for outdoor-parking monitoring and has triggered exports to South American and Eastern European markets

3 Magnets Go Deep Inside: Geodynamic Sensing

In 2009, the World Health Organisation estimated that 3–14% of all cases of lung cancer could be linked to radon gas coming from the inside the Earth! This issue affects not only those people whose occupations have the potential for high exposure to radon, such as mine workers and workers remediating sites contaminated by radioactivity, but also the many millions who reside in areas with high levels of natural radiation, such as may be found in association with uraniferous metamorphic rocks and granites. In the US, sadly, radon has made it to second place, after smoking, as the cause of lung cancer.

Geodynamic activity is seemingly a principal actor in radon outgassing. To address the issues of setting up a world-wide early warning network, the attainment of an automated and accurate sub-micron measurement of geodynamic displacements is required. In 2014, the Czech Republic Grant Agency funded a project for the automatization of a global network of fault- displacement devices as part of a joint cooperation with the Institute of Rock Mechanics of the Czech Academy of Sciences (Dr. Matt Rowberry was named principal investigator) and IGSresearch Ltd.

Several technologies has been explored across several projects; Moiré interference patterns and magnetic positioning were named the two finalists in the validation process. Besides the currently operating international network of co-operating research institutions covering all the continents with data collection instruments, the core team has recently welcomed new academic actors such as IMDEA Nanoscience (Dr. Alberto Bollero as principal investigator) to contribute their protracted experience in permanent magnets and magnetic-field detection (Fig. 3).

Fig. 3 Automated device for in situ analysis of Moiré patterns developed by Institute of Rock Mechanics of the Czech Academy of Sciences and IGSresearch Ltd. The high-accuracy of Moiré interference assessment of displacements has been combined with magnetic-based instrumentation developed jointly with IMDEA Nanoscience

Fig. 4 A demonstration device designed by Dr. Ignasi Fina conceived to explain the piezoelectricity to a broad audience. The device has attracted the attention of the ESCOLAB network and will be replicated and sent to numerous secondary schools during 2015 and 2016

4 Piezoelectrics: Sensing Under Pressure

The flawless ability of magnetic sensing to count cars can be further exploited to assist with wheel counting. With this information at hand, blurring ambiguities in traffic classification are removed. A clever geometric arrangement of carefully selected piezoelectric sensors lead by Prof. Gustau Catalan resulted in a successful engagement between ICN.cat and IGSresearch Ltd.—and a working prototype design to be validated during 2015 and 2016. The cooperation extends beyond the manufacturing of sensors and comprises joint publications in high-impact factor journals and in science outreach too, a field with an intriguingly high potential in the coming years (Fig. 4).

Application of Nanomaterials in Nanomedicine, Dermatology and Cosmetics by Endor Nanotechnologies

Julian Peña Dopazo

1 Introduction

Endor Nanotechnologies is a company founded in 2007 by a former graduate student in physics at the Universidad Autònoma de Barcelona. In the very early stages of the company, the founder shared facilities with the existing research groups at the university, including microscopes, other lab equipment, even computers and the table that 'belonged' to the company was of common use.

So Endor Nanotechnologies had to "spin off" from that initial stage and look for a proper and unshared space to grow. During 2009, the Barcelona Science Park offered an active company-incubator: lab spaces and offices ready to use for biotech and nanotech spin-off and start-up companies, with advantageous conditions during the first four years (basically, lower prices for lab renting and access to the advanced equipment at the Science Park), so Endor Nanotechnologies (Fig. 1) was a perfect candidate to fit that programme.

When finally independent in all senses, and having a transfer agreement with the University to develop a specific technology based on magnetic-metallic nanoparticles, Endor Nanotechnologies began research to addt market applications to that know-how. We quickly completed the protocol for the systematic manufacture of the nanomaterials, and gained the skills to characterize it, meaning to make sure (which is not visible to the naked eye) we were producing nanoparticles (<100 nm) and not any aggregated or bulk material useless for nanotech applications. In Fig. 2

The original version of this chapter was revised: Incorrect volume editor name has been updated. The erratum to this chapter is available at 10.1007/978-3-319-61237-9_12

J. Peña Dopazo (✉)
Endor Nanotechnologies, Parc Científic de Barcelona, Barcelona, Spain
e-mail: julian.p.dopazo@gmail.com

J. Peña Dopazo and F. Zivic (eds.), *Supporting University Ventures in Nanotechnology, Biomaterials and Magnetic Sensing Applications*,
DOI 10.1007/978-3-319-61237-9_9

Fig. 1 Endor Nanotechnologies logo and laboratory facilities. *Image* Endor Nanotechnologies

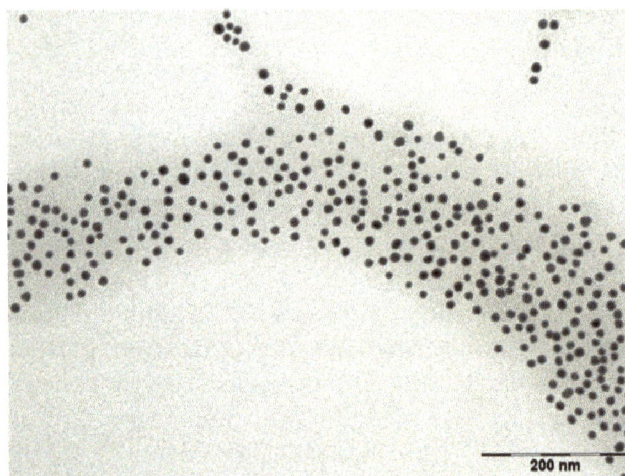

Fig. 2 Hyaluronic acid conjugated to gold nanoparticles. *Image* Endor Nanotechnologies

we can see spherical gold nanoparticles with a size of 10 nm surrounded by hyaluronic acid -observed as a grey shadow- for cosmetic use. Nanoparticles are not aggregated to each another, being this a main feature to keep the special properties of the nanomaterial.

Endor's team consist of a multidisciplinary staff from various technical areas. For scientific performance, there are experts in: chemical engineering, biotechnology and cell biology, most of them with a Ph.D. degree. For management administration, there are business management graduates. Endor's CEO is a graduate in physics with an additional business background.

2 Nanotechnology Products and Services

Endor Nanotechnologies conducts activities within two fields of research: (1) nanomedicine and (2) dermatology and cosmetics. Nanomedicine comprises long-term projects, whereas nanotechnology solutions are already applied in dermatology and

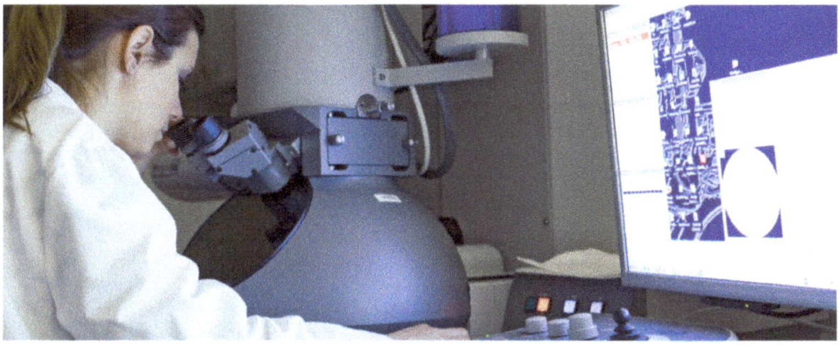

Fig. 3 Observing nanocrystal formation using TEM. *Image* Endor Nanotechnologies

Fig. 4 Hyaluronic acid seen via SEM imaging. *Image* Endor Nanotechnologies

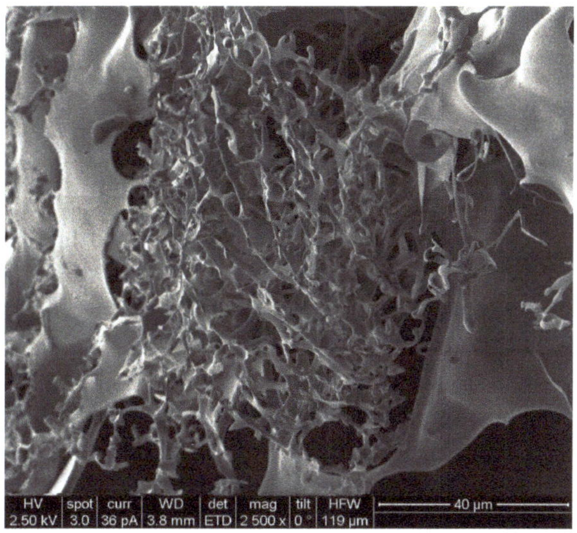

cosmetics. The most advanced technical equipment at the moment is required to enable manufacture and analysis of matter at the nanometric scale. Transmission Electron Microscopy (TEM, Fig. 3) and Scanning Electron Microscopy (SEM, Fig. 4) enable imaging at this range. In Fig. 4, the three-dimensional structure of the image and the deep- field effect obtained from a hyaluronic acid sample is clearly evident.

We develop R&D programs in the oncology area using the properties of our nanoparticles for highly-focused delivery of antitumoural compounds straight to cancer cells, constituting a substantial difference from previous traditional treatments: tailored drugs. Gold nanoparticles are the carriers for the compound we decide to conjugate it with and which we adhere to its surface, so we can test brand new combinations of drugs in a very specific area. In this field, we must comply with the same regulatory criteria as for a regular drug, so it can take up to ten years to get to a final market indication for the product. In Fig. 5, gold nanoparticles have passed

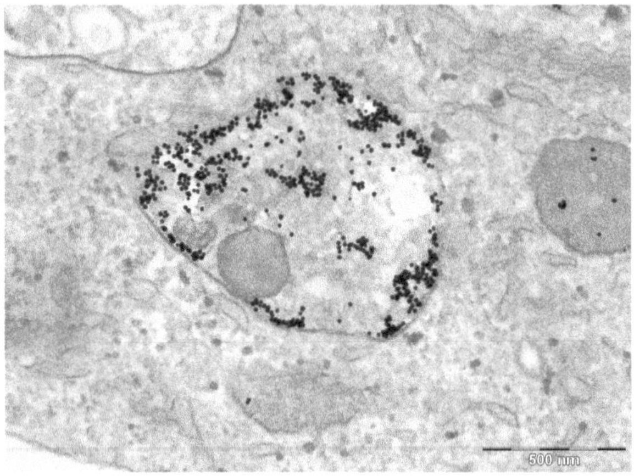

Fig. 5 Gold nanoparticles gathered at a specific cellular endosome. *Image* Endor Nanotechnologies

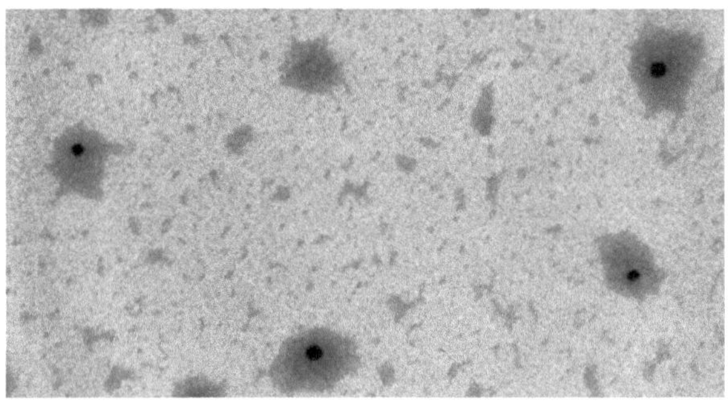

Fig. 6 Isolated nanoparticles wrapped in hyaluronic acid. *Image* Endor Nanotechnologies

through the cell membrane and have fixed onto the targeted endosome. In a real model with tumours, the aim is to target the nanoparticles charged with an active drug precisely where the cancer cells lie. This enables concentration of the treatment and avoids the undesirable side effects of traditional chemotherapy and radiology.

2.1 *Short-Term Projects; Dermatology/Cosmetics*

Endor Nanotechnologies use the 'Golden Hyaluronan' compound (international patent), shown in Fig. 6, as the active ingredient with special efficacy for skin regeneration, to create very close-to-market applications in dermatology and cosmetics. We launched our own cosmetics brand and online shop (YenEffect™), during the fourth quarter of 2012, as the first company in the world to develop a singular ingredient made of nanoparticles for a cosmetic application.

Fig. 7 Yen-Effect™ Serum and cream, containing active nanoparticles. *Image* Endor Nanotechnologies, www.yen-effect.com/en

The result is an extremely effective product that helps the skin to recover from the natural, ongoing decrease of hyaluronic acid that causes skin to lose its firmness, and return the tissue to a more balanced condition. And we think that such an extraordinary product is worth an extraordinary brand image, as shown in Fig. 7.

3 The Future

Endor Nanotechnologies| strategic plan predicts that the relevant financial resources for the oncologic development research over the next years will flow from the short-term projects‘ revenues. The company's original financial resources to begin its activities, apart from the founder's seed capital, came from the 'Centro de

Desarrollo Tecnológico industrial—CDTI' which is the most important national center for funding innovation with loan mechanisms, plus specific grants to hire technical staff for research projects from the national government of Spain. Once our technological protocols were confirmed, the company was able to participate in a FP7 project with a grant contribution from the EU. Endor Nanotechnologies began to generate incomes from nanoparticles sales to the market in 2009, and the plan is to boost this business activity with steady and endurable growth. The main strategy in this field is based on commercial-network implementation and distribution agreements for the most profitable territories for our products: the European Union, the U.S.A. and East Asia.

Screvo: 3D Screening Devices for In Vitro and In Vivo Applications

Julian Higuera

SCREVO

1 Introduction

Screvo BV is an early-stage biotech company founded in 2012 whose mission is to optimize the number of experimental animals in preclinical research. It is a spin-off from the Tissue Regeneration department of the MIRA Institute for Biomedical Technology of the University Twente, The Netherlands. Screvo's first product, *Vers3D*, is an implantable well system that can be used in a wide variety of in vitro and in vivo applications. By increasing the number of conditions simultaneously screened in one animal, animal testing of drugs and treatments is at least three times more time-efficient and at least nine times more cost-effective thanks to this exciting new product.

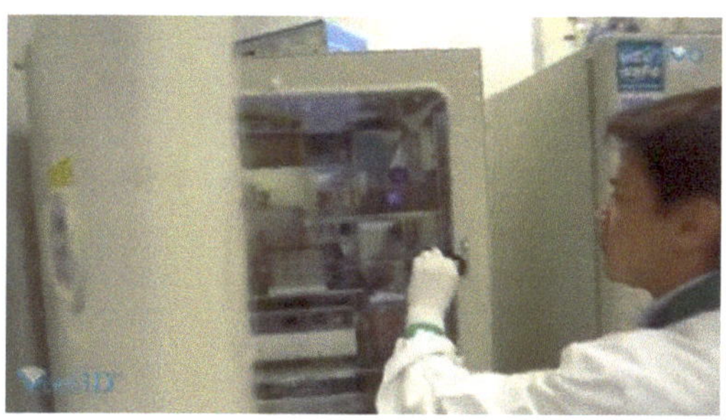

J. Higuera (✉)
Screvo, Enschede, The Netherlands
e-mail: julian.higuera@screvo.net

J. Peña Dopazo and F. Zivic (eds.), *Supporting University Ventures in Nanotechnology, Biomaterials and Magnetic Sensing Applications*,
DOI 10.1007/978-3-319-61237-9_10

2 The Innovation

Back in 2010, while Gustavo was at the end of his PhD studies in the field of tissue regeneration, he was faced with the challenge of answering one fundamental question: "*Could different amounts of stem-cells create a different type of tissue*". Because it was not realistic to thoroughly answer this question with the standard techniques, he developed an animal-implantable screening device that could contain multiple experiments in a single animal. By placing different stem-cells concentrations in a single animal, as opposed to testing just one concentration per animal, he created a screening device capable of carrying up to 36 experimental conditions per animal. In this manner, there could be ninefold savings with respect to the number of animals usually needed to complete such experiments. It was then realized that this new technology could prove to be highly beneficial in the field of pre-clinical research if available to the market.

Screvo's 3D systems can be organized in a matrix array and are designed and fabricated by rapid-prototyping technologies, among others, with biocompatible biomaterials. Furthermore, Screvo's technology can also be integrated within current systems, such as automatic confocal-light-microscopy equipment.

The technology can be used for in vitro and in vivo experiments in scientific research, such as cell culture experiments. The 3D technology platform forms the basis for our approach to product development, and can be used within multiple product variants to address diverse market applications. This last characteristic is both a challenge and an advantage because Vers3D has also been unsuccessfully tested in a broad number of applications already including: stent development and cancer-cell culture in the quest to find its niche for growth, but no start-up can indefinitely test applications. Therefore, we are currently focused on one successful application that holds the highest market promise: the culture of 3D brain-tissue units in Vers3D wells and its drug-screening model potential.

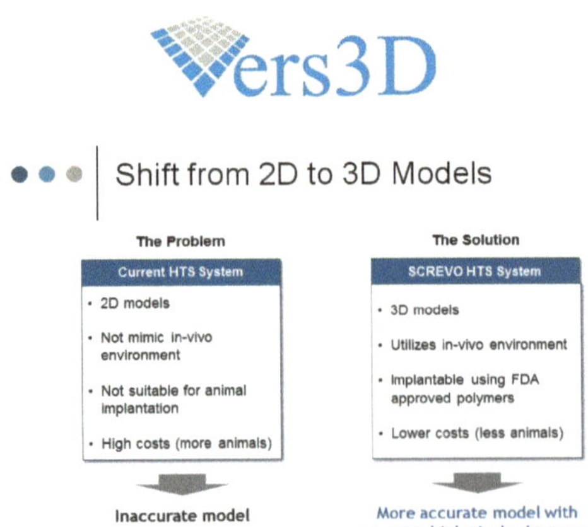

3 Building the Team

Gustavo combined efforts with the MIRA Institute for Biomedical Technology of the University Twente into create Screvo, a spinoff with the purpose of revolutionizing the market for screening devices. Screvo is located in the laboratories of MIRA in Enschede, The Netherlands, and uses state-of-the art facilities for R&D and manufacturing purposes. In addition to this, in order to bring products to market, Gustavo recognized that even though he had ample experience in the field of biological research, he lacked an understanding of how to manage a start-up business, specifically, that he lacked business contacts, did not have a clear business model and had no available funding to take his idea to market. In the fall of 2011, Gustavo devoted efforts to finding a business developer or a group of business developers to help him handle the strategy, marketing and financial portion of the business.

He knew that having a great idea or product did not guarantee great results: if the company were to have any success, Screvo needed the right mix of human capital and financial resources to take the market by storm. Gustavo interacted with various groups in The Netherlands but was not able find a team that comprehended his vision for the company. The people he met, either lacked solid prior business experience in the field of biotechnology or a proven track record of managing a company in a financially effective manner. He needed to find someone who had such experience and who would find taking his idea to market as a personal challenge…

Family always comes first… Gustavo reached out to his brother: that would be me, Julian. In 2009, I had graduated with an MBA in Finance from Tulane University in New Orleans and for the previous five years had been working in the financial industry conducting financial analysis and helping businesses obtain capital in the credit markets. I mentioned to Gustavo that during my MBA studies, I met Stan Eapen who had held several positions in the pharmaceutical industry, with companies such as Schering-Plough, Biogen Idec and Genentech. Gustavo presented us with his vision for Screvo, and Stan and I became partners and business developers for Screvo.

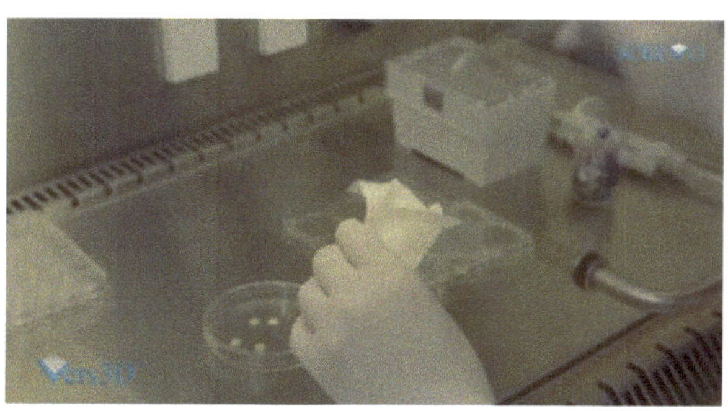

4 Building a Strong Financial Foundation

During the past years, Screvo has overcome numerous challenges. We have made many critical decisions that have helped us bring the company to where it is today, but not without making quite a few mistakes along the way. Looking back, these 'mistakes' have helped us see what works and what does not. One key aspect of starting a company is trying to manage expenses and make the most of the limited available resources. Screvo has learned to recognise what actions are strictly necessary to build a solid company that can function properly, especially during the pre-revenue periods.

The following is a list of focus points concerning the making of sound financial decisions when embarking on starting up a company in today's ever-changing environment:

- **Target available grants and subsidies**: Options may vary depending on where your company is established or the specific industry it is in, but having a backup plan to support the company until initial sales are achieved is key to starting on the right footing.
- **Establish key relationship that will enable you to save money in the initial stages**: Legal and accounting fees may become increasingly high if not managed correctly. Try to contact your local small-business support office to look for pro-bono support in these areas when structuring a company or when it is time to close the books.
- **Attend conferences or trade shows that offer discounts to new companies**: Some conferences even have a limited number of free spaces available for companies who are in initial stages and who are promoting a new exciting technology or product.
- **Do not overspend**: Limit your expenses to only what is really necessary to operate the business. Equipment and technological resources can be costly and sometimes redundant, given a current lack of products/services demand. Start advertising campaigns by using free newsletter applications or low-cost advertising tools.
- **Know who your customers are**: By employing resources to target a specific niche market, marketing dollars can be spent more efficiently and geared towards a concrete area of the market.
- **Hiring talent**: Look for ways to increase headcount only when strictly necessary and there is the business volume to support it. In some instances, when attracting key individuals, offer fractional ownership as a compensation option.

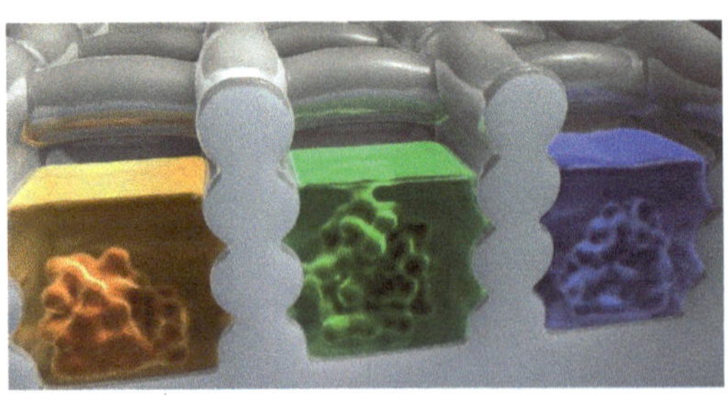

5 Business Development Phase

The business-development phase is a period that needs very clear planning. By allowing a business- development team to provide guidance regarding how to perform a feasibility study at an early stage, Screvo has been able to deliver a clear value-proposition formulation. The company has focused its business-development efforts on creating a *Brand Plan* that identifies the specific ways to create value for the company and includes a *Tactical Plan* for going to market. Screvo has utilized the following step- by-step approach:

Step 1 Demand and Trend Analysis—To validate if there is current demand for 3D screening devices (Vers3D). Conducted a market analysis including a product-quality study and a survey of market size and value.

Step 2 Competitive Assessment—Who else is producing similar products? Confirmed the number and type of competitors and the specific market size.

Step 3 Perform a comprehensive SWOT analysis.

Step 4 Operating Requirements—Identify the physical and fiscal resources required, including production-cost estimates.

Step 5 Business-Plan Elaboration—Plan a commercialisation strategy and long-term development for the business to include the following:

- Value-proposition formulation
- Brand Plan and Tactical Plan
- Market presence and lead-generation strategy
- Financial Projections: Provide profit & loss Statement, balance sheet, cash flow and CAPEX.
- Commercial operations plan (sales-funnel generation and commercial partnerships)
- Market Introduction.

Screvo now plans to scale-up marketing and focus extensively on sales growth. The main objectives for the next three years are:

- Increase revenue via fee-based services.
- Develop licensing deals with potential customers.
- Explore strategic partnership opportunities.
- Raise dilutive and non-dilutive funding.

6 Evolution and Milestones

Since its foundation, Screvo has achieved the following milestones:

- Validation of technology platform (in vivo).
- Acquired 250.000 EUR for the development of the company.

- Secured solid IP position.
- Reached collaboration and patent-transfer agreement with MIRA of the University of Twente.
- Established a solid management team consisting of both scientific and business expertise.
- Beta-customer of well system.
- Performed market research.
- Formulated a Brand Plan and Value Proposition.
- Optimized manufacturing processes.
- Attended the 2012 TERMIS World Congress in Vienna.
- Awarded a slot in the “Innovation AveNew” in January 2013 by the Society for Laboratory Automation and Screening (SLAS).
- Gustavo presented “Smaller is Better” at the 2014 **TEDxRSM** event that took place in Rotterdam.

Our technology is on the right track to make a key impact in the field of preclinical research and our scientific team is confident about Vers3D’s potential. Our vision is that our devices become standard tools for researchers and pharmaceutical companies in optimizing the development of new alternatives to fight diseases.

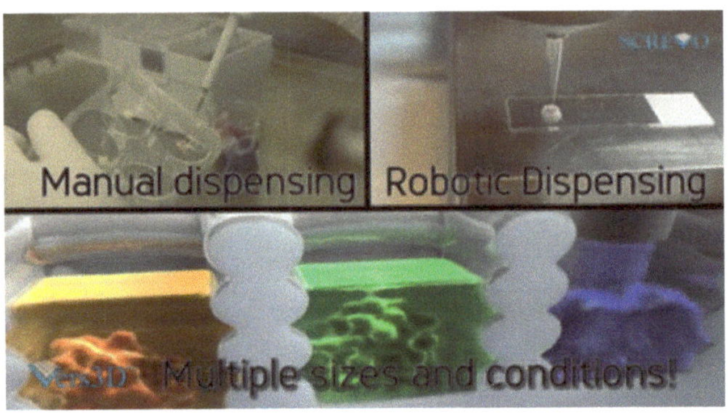

Case Study of Successful Collaboration Between Research and Business: Commercialisation of Advanced Motion-Control Components

Petar Todorović, Branko Tadić and Mina Vasković

1 Introduction

The research process through defined activities that lead to goal achievement usually consists of a rather common set of activities. If the project plan includes an upgrade or development of a new technical solution or a product, then the process begins with analysis of existing solutions and a review of possible directions for upgrade, then continuing on to realization of a new idea.

These steps were also included in realization of our project at the Faculty of Engineering in Kragujevac, Serbia, at the beginning of 2000. Our main goal was to examine the possibility of development and production of a multipurpose CNC machine that would be presented to small/and medium-sized domestic enterprises. After reviewing the actual requirements, it was determined that it be a light-weight CNC milling machine—router, which would be mainly used for flattened material (MDF sheet materials, plastic), but also for wood, soft metals and similar materials. During 2003, this project entered into a final phase of realization with a designed and produced prototype of a light CNC router. However, it was then clear that the available funds would not enable finalization of router production. The cost of control electronics for the router exceeded the available funds. It has to be mentioned that funds for this purpose were significant. This moment in research and analysis was the initial point for our group to consider the current situation as a development option, that is, to begin the applied research and development (R&D) of power electronics for numerical controlled machines.

P. Todorović (✉) · B. Tadić
Department for Production Engineering, Faculty of Engineering, University of Kragujevac, Sestre Janjić 6, 34000 Kragujevac, Serbia
e-mail: petar@kg.ac.rs

M. Vasković
Department of Applied Mechanics and Automation Control, Faculty of Engineering, University of Kragujevac, Sestre Janjić 6, 34000 Kragujevac, Serbia

J. Peña Dopazo and F. Zivic (eds.), *Supporting University Ventures in Nanotechnology, Biomaterials and Magnetic Sensing Applications*, DOI 10.1007/978-3-319-61237-9_11

2 Motion Control Systems

The global market offers various systems for electronic devices for motion control, based on different technologies. Mostly it depends on the actuators, i.e. motors; thus motion control systems could be roughly categorized into three groups, for the sake of simplicity.

One of the first choices are control systems based on stepper motors. At first glance, these systems are the easiest to control and, in most cases, the cheapest solution. Control of stepper motors is mostly realized in open-loop systems, where the position accuracy has been provided with the proper torque sizing for the selected stepper motor. These systems, in some cases, contain a position-feedback system, which make them lose their main comparative advantage—the overall cost.

The second group of control electronic for motion control is based on the application of DC motors with permanent magnets. A DC motor has a simple mathematical model, which makes it easy to control. For precise positioning systems, use of an incremental encoder is mandatory. It gives the angular position to the system, making its control more complicated. DC motor controllers have pre-installed PID regulators, so the final user should be acquainted with some basic settings of the mechanical-system response, which includes the drive engine—the moveable mechanical system on the machine.

The third group of drive electronics is based on the application of AC servo motors and connected control electronics. These systems provide the optimal performance control. At the same time, they are the most expensive systems, if compared with the similar systems from the first and the second group, making them less available to a wide group of end users.

An important contribution to the development of modern motion control systems was made due to three major technological breakthroughs. Some of the most important elements are permanent magnets, which are parts of every motor in all three of the major drive-system groups. The first breakthrough appeared in the 1980s, and it was the discovery of the rare- earth magnets. The rare-earth magnets are permanent magnets that are made with an alloy of rare-earth elements. They are among the strongest magnets ever made. Rare-earth magnets have a magnetic field of around 1.4 T, while ferrite and ceramic permanent magnets produce fields in range of 0.5–1 T. Neodymium and samarium-cobalt magnets are commonly used. Motors which use permanent magnets based on rare-earth elements have significantly better rated power/volume ratio than motors that use ferrite or ceramic permanent magnets.

The next key contribution to the further breakthrough in motion control system domain was the development of new technologies for producing solid-state switching circuitry. Primarily, this refers to the powerful MOSFET and IGBT transistors and power modules based on the described technologies, which have the following characteristics: high-power ratings, high- switching frequency, low-switching losses etc.

The development of microcontrollers was an important impetus for the development of motion control system. A microcontroller is composed of a processor, memory and peripherals within a single chip, which is used for embedded systems. Microcontrollers have enabled implementation of new algorithms for motion-control systems.

These facts and innovations have led to a significant price-drop of motion-control systems, therefore making them available to a broader group of consumers.

3 Our Story

The motivation for the development of motion-control systems was the lack of funds for the already described groups of products in the final project stage. Figure 1 presents subsequent research and the development of motion-control systems.

From 2003 until 2011, difficult and intensive work on R&D of the motion-control system, as well as production and testing of the first prototypes, characterised the project. At the beginning, stepper motors were used; after that, DC servo motors were the base of the control system. In 2007, a drive for stepper motors MST-107 was developed, which was, after a few modifications, still being used as the base of our motion-control systems.

The turning point happened in 2011, when the Prizma company (www.prizma.rs) from Kragujevac, Serbia, started full cooperation with our group, taking over production and distribution of the final products. Prizma d.o.o. is a privately owned company (Fig. 2), and it has developed business activities in two directions: distribution of well-known brands of equipment, and, alongside that, they also realise their own research and development of products in various fields, such as medical electronics, industrial electronics, consumer electronics and automotive industry.

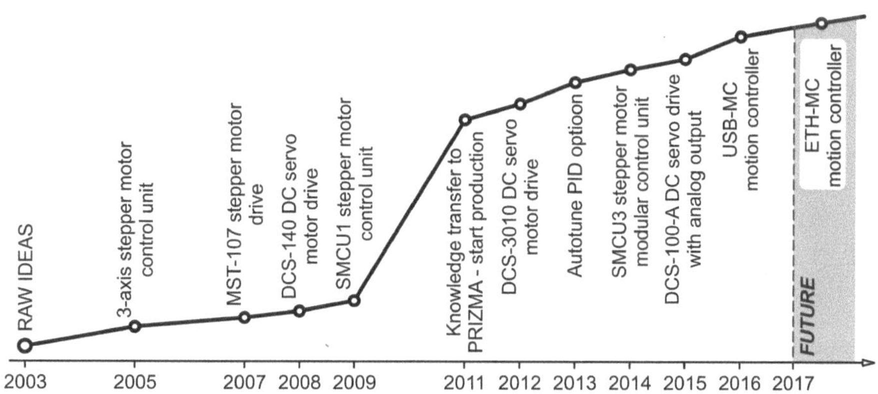

Fig. 1 Motion control systems R&D—our story; detailed product description can be found at http://www.audiohms.com/en/ and http://www.prizma.rs/ml/index.php/motion-controls

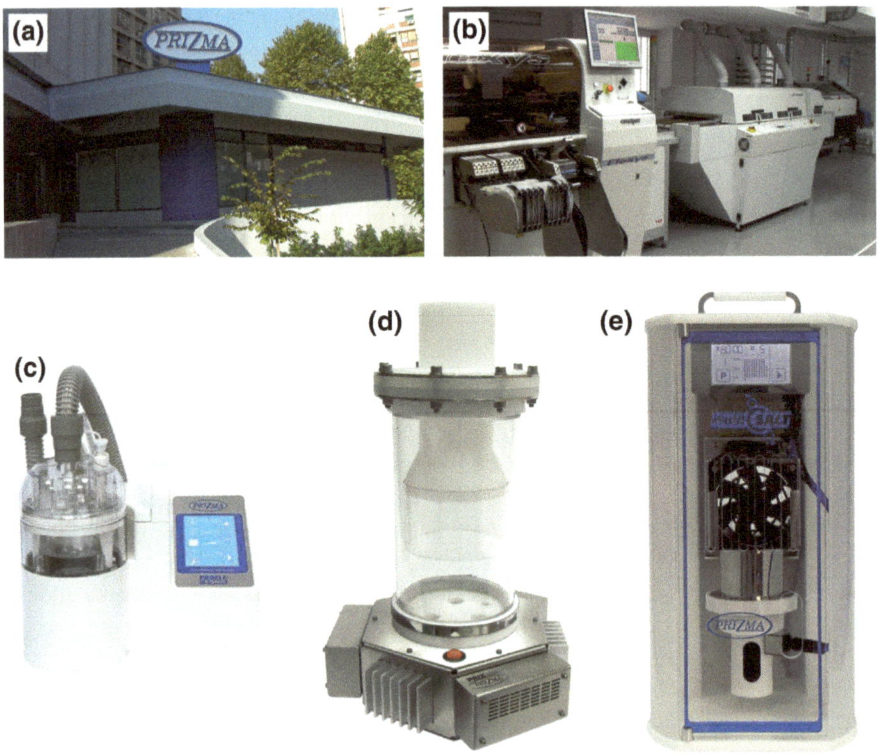

Fig. 2 Prizma doo, **a** facility, **b** production line, **c** Prizma Ultrasonic Nebuliser PROFI SONIC H, **d** Prizma Ultrasonic Pyrolysis Atomizer PRIZnano, and **e** Prizma Dry Salt Aerosol Generator MultiSALT

Cooperation with Prizma started with adaptation of existing product documentation and existing product solutions for serial production. From that moment, a large expansion and launching of the new line of products started. The most recently introduced product is the USB-MC motion controller with extensions, the first product in the motion-controller group. Development of a novel Ethernet-based motion controller (ETH-MC) is in progress, and it is designed to follow the modern trend of transferring all activities in the field of automation to internet-based solutions, such as remote virtual labs.

The four stages of the years of research development of the micro-step drive for stepper motors is shown in Fig. 3, depicting the product "growth" over the years.

All products listed here are embedded systems based on microcontrollers. They have been recognized by a wide range of users who are hobbyists, as well as by professional manufacturers of light CNC machines and other categories of machines that require precise positioning. In addition, service engineers who work in retrofitting of older CNC machines also have become interested. Our motion control products are also implemented in various educational systems based on

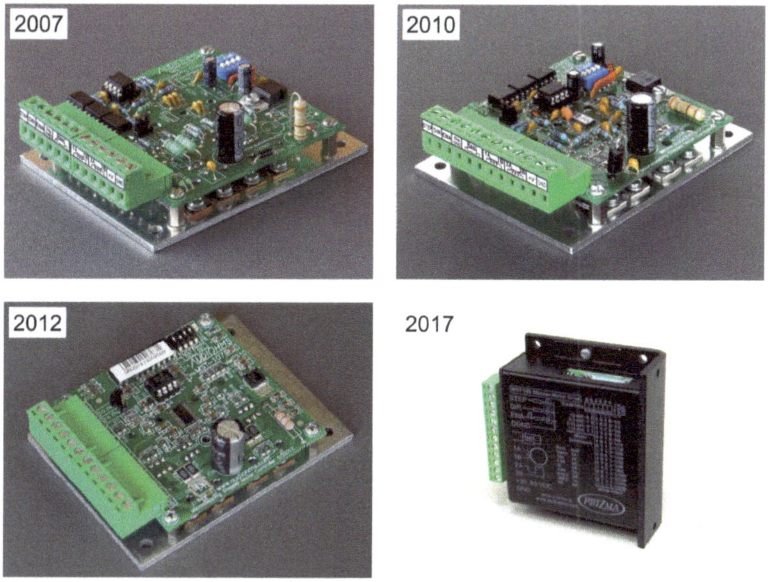

Fig. 3 Stepper motor drive through the years

robots/manipulators, 3D scanners, and 3D printers, thus making our research broadly applied to both commercial products and teaching resources at universities.

For the control of a complete system, open source and commercial software is used, like Mach3 (http://www.machsupport.com/) or LinuxCNC (http://linuxcnc.org/). They transform a basic computer into a CNC control unit. Also, our motion control systems can be used with PLCs or as stand-alone components for users who develop their own embedded or PC-based systems for control.

4 Conclusion

Difficulties in project realization, or in the idea in general, do not have to mean the end of that project. In some cases, problems can become a challenge and lead to development of new ideas which can be successfully materialised. Satisfaction that is derived from the successfully implemented ideas in a new product used in any practical application, especially after problems in its realisation, inevitably produce high self-affirmation and are the best possible outcomes of any research effort.

Erratum to: Supporting University Ventures in Nanotechnology, Biomaterials and Magnetic Sensing Applications

Julian Peña Dopazo and Fatima Zivic

Erratum to:
J. Peña Dopazo and F. Zivic (eds.), *Supporting University Ventures in Nanotechnology, Biomaterials and Magnetic Sensing Applications*, DOI 10.1007/978-3-319-61237-9

In the original version of the book, the incorrect volume editor name "Julian Peña" in Cover page, FM (Title page, Copyright page, Preface and ToC), and Chapter 9 has to be changed to read as "Julian Peña Dopazo". The erratum book has been updated with the change.

The updated online version of this book can be found at
http://dx.doi.org/10.1007/978-3-319-61237-9

J. Peña Dopazo and F. Zivic (eds.), *Supporting University Ventures in Nanotechnology, Biomaterials and Magnetic Sensing Applications*,
DOI 10.1007/978-3-319-61237-9_12

FSC
www.fsc.org
MIX
Papier | Fördert
gute Waldnutzung
FSC® C083411

Kolibri 360 GmbH
Ferdinand-Jühlke-Straße 7
99095 Erfurt, Deutschland
produktsicherheit@kolibri360.de